Data Acquisition Techniques Using Personal Computers

Data Acquisition Techniques Using Personal Computers

Howard Austerlitz
CYBEX
A Division of Lumex, Inc.
Ronkonkoma, New York

Academic Press, Inc.
Harcourt Brace Jovanovich, Publishers
San Diego   New York   Boston
London   Sydney   Tokyo   Toronto

Front cover photograph courtesy of International Business Machines Corporation.

This book is printed on acid-free paper. ∞

Copyright © 1991 by ACADEMIC PRESS, INC.
All Rights Reserved.
No part of this publication may be reproduced or transmitted in any form or by any means, electronic or mechanical, including photocopy, recording, or any information storage and retrieval system, without permission in writing from the publisher.

IBM PC, PC/XT, PC/AT, PS/2, PC DOS and Micro Channel are trademarks of IBM Corporation. MS DOS, Microsoft Windows and Microsoft C are trademarks of Microsoft Corporation. OS/2 is a trademark of Microsoft and IBM Corporation. Unix is a trademark of AT&T Information Systems. Apple Macintosh is a trademark of Apple Computer, Inc. Nu Bus is a trademark of Texas Instruments. Intel is a trademark of Intel Corporation. Motorola is a trademark of Motorola Corporation. Lotus 1-2-3 and Symphony are trademarks of Lotus Corporation. ASYST is a trademark of Keithley Asyst. LABTECH NOTEBOOK is a trademark of Laboratory Technologies Corporation.

Academic Press, Inc.
San Diego, California 92101

United Kingdom Edition published by
Academic Press Limited
24–28 Oval Road, London NW1 7DX

Library of Congress Cataloging-in-Publication Data

Austerlitz, Howard.
    Data acquisition techniques using personal computers / Howard
  Austerlitz.
      p. cm.
    Includes bibliographical references and index.
    ISBN 0-12-068370-9
    1. Microcomputers.   2. Automatic data collection systems.
3. Computer interfaces.   I. Title.
TK7888.3.A872   1991
004.6'16--dc20   91-2754
                                                                         CIP

PRINTED IN THE UNITED STATES OF AMERICA
91 92 93 94   9 8 7 6 5 4 3 2 1

*To my wife Kiel,
whose guidance and understanding
made it all possible*

Contents

Preface xi

CHAPTER **1**
Introduction to Data Acquisition

CHAPTER **2**
Analog Signal Transducers

2.1 Temperature Sensors 7
2.2 Optical Sensors 8
2.3 Force and Pressure Transducers 12
2.4 Magnetic Field Sensors 15
2.5 Ionizing Radiation Sensors 17
2.6 Position (Displacement) Sensors 19
2.7 Humidity Sensors 22
2.8 Fluid Flow Sensors 23

CHAPTER **3**
Analog Signal Conditioning

3.1 Signal Conditioning Techniques 24
3.2 Analog Circuit Components 25
3.3 Analog Conditioning Circuits 31

CHAPTER **4**
Analog/Digital Conversions

4.1 Digital Quantities 40
4.2 Data Conversion and DACs 44
4.3 ADCs 50

viii Contents

CHAPTER 5
The Personal Computer

5.1 IBM PC/XT/AT and Compatible Computers   65
5.2 The IBM PC/XT   66
5.3 The IBM PC/AT   73
5.4 The BIOS   78
5.5 PC Peripherals   78

CHAPTER 6
Interfacing Hardware to the PC Bus

6.1 I/O Data Transfers   83
6.2 Memory Data Transfers   85
6.3 A Simple 8-Bit I/O Port Design   86
6.4 DMA   89
6.5 Wait State Generation   90
6.6 Analog Input Card Design   91
6.7 16-Bit Data Transfers on ISA Computers   92

CHAPTER 7
Interfacing Software to the PC

7.1 PC Software Layers   95
7.2 Software Interrupts   97
7.3 Polled versus Interrupt-Driven Software   100
7.4 Device Drivers   104
7.5 TSR Programs   104
7.6 DOS   105
7.7 Non-DOS Operating Systems and Software Environments   106
7.8 Overcoming DOS Memory Limitations   107
7.9 Software Support for a Mouse   110

CHAPTER 8
Standard Hardware Interfaces

8.1 Parallel versus Serial Digital Interfaces   112
8.2 Parallel Interfaces   114
8.3 Serial Interfaces   127

CHAPTER 9
Data Storage and Compression Techniques

9.1 DOS Disk Structure and Files   143
9.2 Common DOS File Types   147
9.3 Data Compression Techniques   151

CHAPTER 10
Data Processing and Analysis

10.1 Numerical Representation   171
10.2 Data Analysis Techniques   178

CHAPTER 11
Commercial Data Acquisition Products

11.1 Commercial Data Acquisition Hardware Products   199
11.2 Commercial Data Acquisition Software Products   216
11.3 How to Choose Commercial Data Acquisition Products   228

CHAPTER 12
Other Personal Computer Systems and Hardware

12.1 IBM PS/2 Personal Computers with MCA   231
12.2 Apple Macintosh II Computers with NuBus   237
12.3 Math Coprocessors   243
12.4 Other Processor Cards   247
12.5 Specialized Personal Computer Systems   248

CHAPTER 13
Computer Programming Languages

13.1 Assembly Language   254
13.2 BASIC   257
13.3 C Programming Language   261
13.4 FORTRAN   265
13.5 Pascal   267
13.6 Considerations for Writing Computer Programs   270

CHAPTER 14
PC-Based Data Acquisition Applications
14.1 Ultrasonic Measurement System 274
14.2 Electrocardiogram (ECG) Measurement System 281
14.3 Commercial Equipment Using Embedded PCs 286
14.4 Future Trends in PC-Based Data Acquisition 292

APPENDIX A
Data Acquisition Hardware Manufacturers 295

APPENDIX B
Data Acquisition Software Manufacturers 301

Bibliography 305
Index 307

Preface

In recent years personal computers (PCs) have become common fixtures in most laboratories due to their low cost and wide range of hardware and software support. They have replaced minicomputers as *de facto* platforms for data acquisition systems. *Data Acquisition Techniques Using Personal Computers* is intended to be a tutorial and reference for engineers, scientists, students, and technicians interested in using personal computers for data acquisition and analysis.

It is assumed that the reader knows the basic workings of personal computers and electronic hardware, although these aspects will be reviewed briefly in this work. Sources listed in the bibliography are good introductions to many of these topics.

Only the family of IBM PCs and compatible systems (PC/XT/AT computers) will be covered in any great detail here, since they represent the largest hardware and software support base for scientific and engineering applications. However, IBM's PS/2 systems (based on Micro Channel) and Apple's Macintosh II computers (based on NuBus) will be covered briefly.

This book stresses "real" applications and includes specific examples as well as a survey of commercially available hardware and software products. It is intended to provide all the information you need to set up a data acquisition system based on a personal computer. In addition, it will serve as a useful reference on personal computer technology.

The area of software is as important as hardware, if not more so. Software topics, such as programming languages, interfacing to a PC's software environment, and data analysis techniques, are covered in detail, along with a survey of commercial data acquisition application programs.

Throughout this work, the term *personal computer* will refer to a generic machine. It can be an Apple Macintosh or an IBM PS/2 system. The abbreviated term *PC* will imply an IBM PC/XT/AT system or compatible, based on an Intel 80x86 family microprocessor and running MS-DOS (or IBM DOS) software.

I wish to acknowledge the many people who helped me with this undertaking. My thanks to Academic Press for getting the project started and seeing it through to its conclusion. I am grateful for the assistance I received from manufacturers in the data acquisition field, including Ved Vasconcelos from Keithley Metrabyte, Kate Kressman from Keithley Asyst, Shari Worthington from Laboratory Technologies, and Iris Polaski from Burr-Brown/Intelligent Instrumentation. I also wish to thank everyone at CYBEX who helped me, especially Jim Smith. Finally, I want to acknowledge Orndorff, the lap-top editor who kept me company during all those late nights at my PC.

Howard Austerlitz

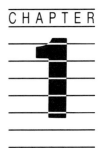

CHAPTER 1

Introduction to Data Acquisition

Data acquisition, in the general sense, is the process of collecting information from the real world. For most engineers and scientists these data are mostly numerical and usually collected, stored, and analyzed using a computer. The use of a computer automates the data acquisition process, enabling the collection of more data in less time with fewer errors. This book deals solely with automated data acquisition using personal computers.

An illustrative example of the utility of automated data acquisition is measuring the temperature of a heated object versus time. Human observers are limited in how fast they can record readings (say, every second, at best) and how much data can be recorded before errors due to fatigue occur (perhaps after 5 minutes or 300 readings). An automated data acquisition system can easily record readings for very small time intervals (i.e., much less than a millisecond), continuing for arbitrarily long time periods (limited mainly by the amount of storage media available). In fact, it is easy to acquire too much data, which can complicate the subsequent analysis. Once the data are stored in a computer, they can be displayed graphically, analyzed, or otherwise manipulated.

Most real-world data are not in a form that can be directly recorded by a computer. These quantities typically include temperature, pressure, distance, velocity, mass, and energy output (such as optical, acoustic, and electrical energy). Very often these quantities are measured versus time or position. A physical quantity must first be converted to an electrical quantity (voltage, current, or resistance) using a *sensor* or *transducer*. This enables the data to be conditioned by electronic instrumentation,

2 CHAPTER 1 Introduction to Data Acquisition

which operates on analog signals or waveforms (a signal or waveform is an electrical parameter, most often a voltage, that varies with time). This analog signal is *continuous* and *monotonic*; that is, its values can vary over a specified range (for example, somewhere between −5.0 volts and +3.2 volts). The values can change an arbitrarily small amount within an arbitrarily small time interval.

To be recorded (and understood) by a computer, data must be in digital form. Digital waveforms have discrete values (only certain values are allowed) and have a specified (usually constant) time interval between values. This gives them a "stepped" (noncontinuous) appearance, as shown by the digitized sawtooth in Figure 1-1. When this time interval becomes small enough, the digital waveform becomes a good approximation of the analog waveform. If the transfer function of the transducer and the analog instrumentation is known, the digital waveform can be an accurate representation of the time-varying quantity to be measured.

The process of converting an analog signal to a digital one is called analog-to-digital conversion, and the device that does this is an analog-to-digital converter (ADC). The resulting digital signal is usually an array of digital values of known range (scale factor) separated by a fixed time interval (or sampling interval). If the values are sampled at irregular time intervals, the acquired data will contain both value and time information.

The reverse process of converting digital data to an analog signal is called digital-to-analog conversion, and the device that does this is called a digital-to-analog converter (DAC). Some common applications for DACs include control systems, waveform generators, and speech synthesizers.

A general purpose laboratory data acquisition system typically consists of ADCs, DACs, and simple digital inputs and outputs. Figure 1-2 is

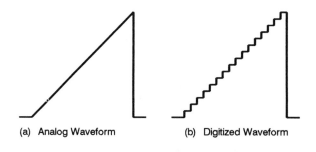

(a) Analog Waveform   (b) Digitized Waveform

Figure 1-1 Comparison of analog and digitized waveforms: (a) sawtooth analog waveform and (b) a coarse digitized representation.

Introduction to Data Acquisition 3

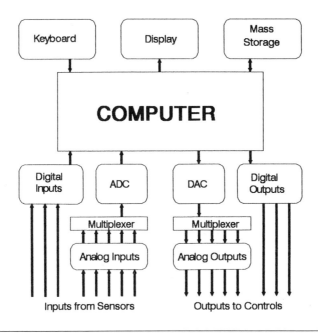

Figure 1-2 Simplified block diagram of a data acquisition system.

a simplified block diagram of such a system. Note that additional channels are often added to an ADC or DAC via a *multiplexer* (or *mux*), used to select which one of the several analog input signals to convert at any given time. This is an economical approach when all the analog signals do not need to be simultaneously monitored.

Economics is a major rationale behind using personal computers for data acquisition systems. The typical data acquisition system of 10–15 years ago, based on a minicomputer, cost about 20 times as much as today's systems, based on personal computers, at around the same performance levels. This is largely due to the continuing decrease of electronic component costs along with increased functionality (more logic elements in the same package). Since personal computers have become commonplace in most labs, the cost of implementing a data acquisition system is often just the price of an add-in board and support software, which is usually a moderate expense.

There are, of course, applications where a data acquisition system based on a personal computer is not appropriate and a more expensive, dedicated system should be used. The important system parameters for making such a decision include sampling speed, accuracy, resolution, amount of data, multitasking capabilities, and the required data processing and display.

Personal computer-based systems have certain limitations in these areas, especially regarding sampling speed and handling large amounts of data. However, newer, high-performance personal computers keep "pushing the edge of the envelope"; they can out-perform dedicated data acquisition systems. The evolution of the PCs based on the Intel 80×86 microprocessor (or CPU) family, the IBM PC/XT/AT, PS/2, and compatible systems, is demonstrated in Table 1-1, showing bus width and the amount of available memory space.

In recent years, Apple's Macintosh computer line has gained popularity as a platform for data acquisition, now that a nonproprietary interface, NuBus, is used. These machines, based on the Motorola 68000 family of microprocessors, have certain advantages, including a graphical, consistent operating environment (using icons) and a linear memory addressing space. (The segmented addressing space of the Intel 80×86 family will be discussed in Chapter 5.)

Software is as important to data acquisition systems as hardware capabilities. Inefficient software can waste the usefulness of the most able data acquisition hardware system. Conversely, well-written software can squeeze the maximum performance out of mediocre hardware. Software selection is at least as important as hardware selection and often more complex.

Data acquisition software controls not only the collection of data but also its analysis and eventual display. Ease of data analysis and presentation are the major reasons behind using computers for data acquisition in the first place. With the appropriate software, computers can process the

TABLE 1-1
INTEL 80×86 CPU Family Bus Width Characteristics

CPU	DATA BUS SIZE (bits)	ADDRESS BUS SIZE (bits)	MEMORY SPACE (Mbytes)
8086	16	20	1
8088	8	20	1
80286	16	24	16
80386	32	32	4096
80486	32	32	4096

acquired data and produce outputs in the form of tables or plots. Without these capabilities, the equipment is not much more than a sophisticated (and expensive) data recorder.

An additional area of software use is that of control. Computer outputs may control some aspects of the system that is being measured, as in automated industrial process controls. The software must be able to measure system parameters, make decisions based on those measurements, and vary the computer outputs accordingly. For example, in a temperature regulation system, the input would be a temperature sensor and the output would control a heater. In control applications, software reliability and response time are paramount. Slow or erroneous software responses could cause physical damage.

A plethora of commercially available PC-based software packages can collect, analyze, and display data graphically, using little or no programming (see Chapter 11). This software allows the user to concentrate on the application instead of worrying about the mechanics of getting data from point A to point B or how to plot a set of Cartesian coordinates. Many commercial software packages contain all three capabilities of data acquisition, analysis, and display (the so-called integrated packages), while others are optimized for only one or two of these areas.

The important point is that you do not have to be a computer expert or even a programmer to implement an entire personal computer-based data acquisition system. Best of all, you do not have to be rich, either.

The next chapter examines the world of analog signals and their transducers, the "front end" of any data acquisition system.

CHAPTER

Analog Signal Transducers

Most real-world events and their measurements are analog. That is, the measurements can take on a wide, nearly continuous range of values. The physical quantities of interest can be as diverse as heat, pressure, light, force, velocity, or position. To be measured using an electronic data acquisition system, these quantities must first be converted to electrical quantities such as voltage, current, or impedance.

A transducer converts one physical quantity into another. For the purposes of this book, all the transducers mentioned convert physical quantities into electrical ones, for use with electronic instrumentation. The mathematical description of what a transducer does is its transfer function, often designated H. So the operation of a transducer can be described as

$$\text{Output quantity} = H \times \text{Input quantity}$$

Since the transducer is the "front end" of the data acquisition system, its properties are critical to the overall system performance. Some of these properties are sensitivity (the efficiency of the energy conversion), stability (output drift with a constant input), noise, dynamic range, and linearity. Very often the transfer function is dependent on the input quantity. It may be a linear function for one range of input values and then become nonlinear for another range (such as a square-law curve). Looking at sensitivity and noise, if the transducer's sensitivity is too low, or its noise level too high, signal conditioning may not produce an adequate signal-to-noise ratio.

Often the transducer is the last consideration in a data acquisition system, since it is considered mundane. Yet, it should be the primary consideration. The characteristics of the transducer in large part determine the limits of a system's performance.

Now we will look at some common transducers in detail.

2.1 Temperature Sensors

Temperature sensors have electrical parameters that vary with temperature, following well-characterized transfer functions. In fact, nearly all electronic components have properties that vary with temperature. Many of them could potentially be temperature transducers, if their transfer functions were well behaved and insensitive to other variables.

2.1.1 Thermocouples

The *thermocouple* converts temperature to a small DC voltage or current. It consists of two dissimilar metal wires in intimate contact in two or more junctions. The output voltage varies linearly with temperature difference between the junctions—the higher the temperature difference, the higher the voltage output. This linearity is a chief advantage of using a thermocouple, as well as its ruggedness as a sensor.

Disadvantages include low output voltage (especially at lower temperatures), susceptibility to noise (both externally induced and internally caused by wire imperfections and impurities), and the need for a reference junction (at a known temperature) for calibration. When several thermocouples made of the same materials are combined in series, they are called a *thermopile*. The output voltage of a thermopile consists of the sum of all the individual thermocouple outputs, resulting in increased sensitivity. All the reference junctions are kept at the same temperature.

2.1.2 Thermistors

A *thermistor* is a temperature-sensitive resistor with a large, nonlinear, negative temperature coefficient. That is, its resistance decreases nonlinearly as temperature increases. It is usually composed of a mixture of semiconductor materials. It is a very sensitive device, but it has to be properly calibrated for the desired temperature ranges. Repeatability from device to device is not very good. Over relatively small temperature ranges it can approximate a linear response. It is prone to self-heating errors due to the power dissipated in it ($P = I^2R$). This effect is minimized by keeping the current passing through the thermistor to a minimum.

2.1.3 Resistance Temperature Detectors

Resistance temperature detectors (RTDs) rely on the temperature dependence of a material's electrical resistance. They are usually made of a pure metal having a small but accurate positive temperature coefficient. The most accurate RTDs are made of platinum wire and are well characterized and linear from 14 K to higher than 600°C.

2.1.4 Monolithic Temperature Transducers

The *monolithic temperature transducer* is a semiconductor temperature sensor combined with all the required signal conditioning circuitry and located in one integrated circuit. This device typically produces an output voltage proportional to the absolute temperature, with very good accuracy and sensitivity (a typical device produces an output of 10 mV per degree kelvin over a temperature range of 0–100 degrees Celsius). The output of this device can usually go directly into an ADC with very little signal conditioning.

2.2 Optical Sensors

Optical sensors are used for detecting light intensity. Typically, they only respond to particular wavelengths or spectral bands. One sensor may respond only to visible light in the blue-green region, while another sensor may have a peak sensitivity to near-infrared radiation.

2.2.1 Vacuum Tube Photosensors

This class of transducers consists of special-purpose vacuum tubes used as optical detectors. They are all relatively large, require a high-voltage power supply to operate, and are only used in specialized applications (as is true with vacuum tubes in general). These sensors exploit the photoelectric effect, when photons of light striking a suitable surface produce free electrons.

The *vacuum photodiode* consists of a photocathode and anode in a glass or quartz tube. The photocathode emits electrons when struck by photons of light. These electrons are accelerated to the anode by the high (+) voltage and produce a current pulse in the external load resistor R_L (see Figure 2-1). These tubes have relatively low sensitivity, but they can detect high-frequency light variations or modulation (as high as 100 MHz to 1 GHz), for an extremely fast response.

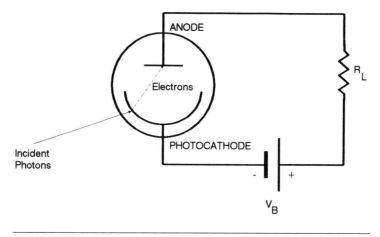

Figure 2-1 Vacuum photodiode.

The *gas photodiode* is similar to a vacuum photodiode, except the tube contains a neutral gas. A single photoelectron (emitted by the photocathode) can collide with several gas atoms, ionizing them and producing several extra electrons. So, more than one electron reaches the anode for every photon. This gas amplification factor is usually 3–5 (larger values cause instabilities). These tubes have a limited frequency response of less than 10 kHz, resulting in a much slower response time.

The *photomultiplier tube* (PMT) is the most popular vacuum tube device in this category. It is similar to a vacuum photodiode with several extra electrodes between the photocathode and anode, called dynodes. Each dynode is held at a greater positive voltage than the previous dynode (and the cathode) via a resistor voltage-divider network (see Figure 2-2). Photoelectrons emitted by the photocathode strike the first dynode, which emits several secondary electrons for each photoelectron, amplifying the photoelectric effect. These secondary electrons strike the next dynode and release more electrons. This process continues until the electrons reach the end of the dynode amplifier chain. There the anode collects all the electrons produced by a single photon, resulting in a relatively large current pulse in the external circuit.

The PMT exhibits very high gain, in the range of 10^5–10^7 electrons emitted per incident photon. This is determined by the number of dynodes, the photocathode sensitivity, power supply voltage, and tube design factors. Some PMTs can detect single photons!

A PMT's output pulses can be measured as a time-averaged current (good for detecting relatively high light levels) or in an individual pulse-

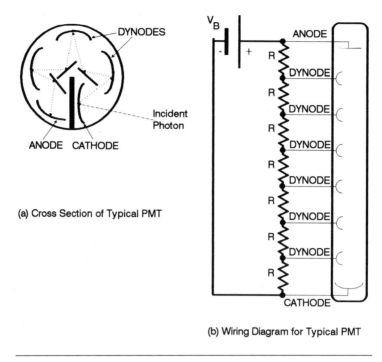

Figure 2-2 Photomultiplier tube (PMT).

counting mode (good for very low light levels) measuring the number of pulses per second. Then a threshold level is used to filter out unwanted pulses (noise) below a selected amplitude.

Some of the noise produced in a PMT is spontaneous emission from the electrodes, which occurs even in the absence of light. This is called the dark count, which determines the PMT's sensitivity threshold. So, the number of photons striking the PMT per unit time must be greater than the dark count for the photons to be detected.

2.2.2 Photoconductive Cells

A photoconductive-cell consists of a thin layer of material such as cadmium sulfide (CdS) or cadmium selenide (CdSe) sandwiched between two electrodes, with a transparent window. The resistance of a cell decreases as the incident light intensity increases. These cells can be used with any resistance-measuring apparatus, such as a bridge. They are commonly used in photographic light meters. A photoconductive cell is usually classified by maximum (dark) resistance, minimum (light) resistance, spectral response, maximum power dissipation, and response time (frequency).

These devices are usually nonlinear and have aging and repeatability problems. They exhibit hysteresis in their response to light. For example, the same cell exposed to the same light source may have a different resistance, depending on the light levels it was previously exposed to.

2.2.3 Photovoltaic (Solar) Cells

These sensors are similar in construction to photoconductive cells. They are made of a semiconductor material, usually silicon (Si) or gallium arsenide (GaAs), which produces a voltage when exposed to light (of suitable wavelength). They require no external power supply, and very large cells can be used as DC power sources. They have a relatively slow response time to light variations but are fairly sensitive. Since the material used must be grown as a single crystal, large photovoltaic cells are very expensive.

A large amount of research has been conducted in recent years in an attempt to produce less expensive photovoltaic cells, made from either amorphous or polycrystalline semiconductors. If these low-cost devices could attain adequate light-conversion efficiency, they would become a practical source of electric energy.

2.2.4 Semiconductor Light Sensors

The members of this class of transducers are all based on a semiconductor device, such as a diode or transistor, whose output current is a function of the light (of suitable wavelength) incident on it.

The *photodiode* is a PN junction diode with a transparent window that produces charge carriers (holes and electrons) at a rate proportional to the incident light intensity. So the photodiode acts as a photoconductive device, varying the current in its external circuit (but, being a semiconductor, it does not obey Ohm's law). A photodiode is a versatile device with a high frequency response and a linear output but low sensitivity, and it usually requires large amounts of amplification.

The *phototransistor* is similar to a photodiode, except that the transistor can provide amplification of the PN junction's light-dependent current. The transistor's emitter-base junction is the light-sensitive element. A *photodarlington* is a special phototransistor, composed of two transistors in a high-gain circuit. The phototransistor offers much higher sensitivity than the photodiode at the expense of a much lower bandwidth (response time) and poorer linearity.

The *charge-coupled device* (CCD) is a special optical sensor consisting of an array (one or two dimensional) of light-sensitive elements. When photons strike a photosensitive area, electron/hole pairs are created in the semiconductor crystal. The holes move into the substrate and the electrons remain in the elements, producing a net electrical charge. The amount of charge is proportional to the amplitude of incident light and the exposure time. The charge at each photosensitive element is then read out serially, via support electronics. CCDs are commonly used in many imaging systems, including video cameras.

2.2.5 Thermoelectric Optical Sensors

This class of transducers convert incident light to heat and produce a temperature output dependent on light intensity, by absorbing all the incident radiation in a "black box." They generally respond to a very broad light spectrum and are relatively insensitive to wavelength, unlike vacuum tube and solid-state sensors. However, they have very slow response times and low sensitivities and are best suited for measuring static or slowly changing light levels, such as calibrating the output of a light source.

The *bolometer* varies its resistance with thermal energy produced by incident radiation. The most common detector element used in a bolometer is a thermistor. They are also commonly used for measuring microwave power levels.

The *thermopile*, as discussed under temperature sensors, is more commonly used than individual thermocouples in light-detecting applications because of its higher sensitivity. It is often used in infrared detectors.

2.3 Force and Pressure Transducers

A wide range of sensors are used for measuring force and pressure. Most pressure transducers rely on the movement of a diaphragm mounted across a pressure differential. The transducer measures this minute movement. Capacitive and inductive pressure sensors operate the same way as capacitive and inductive displacement sensors, which are described later on.

2.3.1 Strain Gages

Strain gages are transducers used for directly measuring forces and their resulting strain on an object. Stress on an object produces a mechanical deformation—strain—defined as

$$\text{Strain} = \frac{\text{length change}}{\text{length}}$$

Strain gages are conductors (often metallic) whose resistance varies with strain. For example, as a wire is stretched, its resistance increases. Strain gages are bonded to the object under stress and are subject to the same forces. They are sensitive to strain in one direction only (the axis of the conductor).

A simple *unbonded* strain gage consists of free wires on supports bonded to the stressed surface. These are not usually used (outside of laboratory demonstrations) because of their large size and mechanical clumsiness.

The *bonded* strain gage overcomes these problems by putting a zigzag pattern of the conductor on an insulating surface, as shown in Figure 2-3. These are relatively small, have good sensitivity, and are easily bonded to the surface under test. The conductor in a bonded strain gage is either a metallic wire, foil, or thin film.

Strain gage materials must have certain well-controlled properties. The most important is sensitivity or gage factor (GF), which is the change in resistance per change in length. Most metallic strain gages have a GF in the range of 2 to 6. The material must also have a low temperature coefficient of resistance as well as stable elastic properties and high tensile strength. Often, strain gages are subject to very large stresses as well as wide temperature swings.

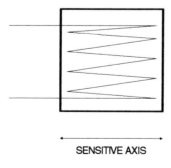

SENSITIVE AXIS

Figure 2-3 Simple, one-dimensional strain gage.

Semiconductor strain gages, usually made of silicon, have a much higher GF than metals (typically in the range of 50 to 200). However, they also have much higher temperature coefficients, which have to be compensated for. They are commonly used in monolithic pressure sensors.

Because of their relatively low sensitivities (resistance changes nominally 0.1 to 1.0%), strain gages require bridge circuits to produce useful outputs. (We will discuss bridge circuits in Chapter 3.) If a second, identical strain gage, not under stress, is put into the bridge circuit, it acts as a temperature compensator.

2.3.2 Piezoelectric Transducers

Piezoelectric transducers are used for, among other things, measuring time-varying forces and pressures. They do not work for static measurements, since they produce no output from a constant force or pressure.

Certain crystalline materials (including quartz, barium titanate, and lithium niobate) generate an electromotive force (emf) when mechanically stressed. Conversely, when a voltage is applied to the crystal, it will become mechanically distorted. This is the piezoelectric effect.

If electrodes are placed on suitable (usually opposite) faces of the crystal, the direction of the deforming force can be controlled. If an AC voltage is applied to the electrodes, the crystal can produce periodic motion, resulting in an acoustic wave, which can be transmitted through other material. When an acoustic wave strikes a piezoelectric crystal it produces an AC voltage.

When a piezoelectric crystal oscillates in the thickness or longitudinal mode, an acoustic wave is produced where the direction of displacement is the direction of wave propagation, as shown in Figure 2-4a. When the crystal's thickness equals a half-wavelength of the longitudinal wave's frequency (or an odd multiple half-wavelength), it is resonant at that frequency. At resonance its mechanical motion is maximum along with the acoustic wave output. And when it is detecting acoustic energy, the output voltage is maximum for the resonant frequency.

This characteristic is applied to quartz crystal oscillators used as highly accurate electronic frequency references in a broad range of equipment, from computers to digital watches.

Typically, piezoelectric crystals are used as ultrasonic transducers for frequencies above 20 kHz, up to about 100 MHz. The limitation on frequency range is due to the impracticalities of producing crystals thin enough for very high frequencies, or the unnecessary expense of producing very thick crystals for low frequencies (where electromagnetic transducers work better).

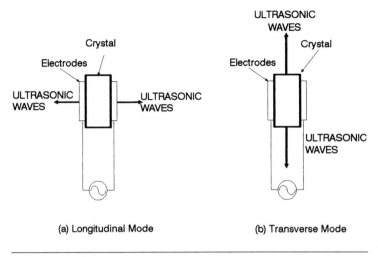

Figure 2-4 Oscillation modes of piezoelectric crystals.

Other crystal deformation modes are transverse, where the direction of motion is at right angles to the direction of wave propagation (as shown in Figure 2-4b), and shear, which is a mix of longitudinal and transverse modes. These modes all have different resonant frequencies.

Piezoelectric transducers have a wide range of applications besides dynamic pressure and force sensing, including

1. Acoustic microscopy for medical and industrial applications, such as "seeing" through materials that are optically opaque. An example is the sonogram.
2. Distance measurements including sonar and range finders.
3. Sound and noise detection such as microphones and loudspeakers for audio and ultrasonic acoustic frequencies.

2.4 Magnetic Field Sensors

This group of transducers is used to measure either varying or fixed magnetic fields.

2.4.1 Varying Magnetic Field Sensors

These transducers are simple inductors (coils) that can measure time-varying magnetic fields such as those produced from an AC source. The magnetic flux through the coil changes with time, so an AC voltage is induced that is proportional to the magnetic field strength.

These devices are often used to measure an alternating current (which is proportional to the AC magnetic field). For standard 60-Hz loads, transformers are used that clamp around a conductor (no direct electrical contact). These are usually low-sensitivity devices, good for alternating currents greater than 0.1 ampere.

2.4.2 Fixed Magnetic Field Sensors

Several types of transducers are commonly used to measure static and slowly varying magnetic fields such as those produced by a permanent magnet or a DC electromagnet.

Hall Effect Sensors When a current-carrying conductor strip is placed with its plane perpendicular to an applied magnetic field (B) and a control current (I_c) is passed through it, a voltage (V_H) is developed across the strip at right angles to I_c and B, as shown in Figure 2-5. V_H is known as the Hall voltage, and this is the Hall effect:

$$V_H = KI_c B/d$$

where

B = magnetic field (in gauss)
d = thickness of strip
K = Hall coefficient

The value of K is very small for most metals but relatively large for certain n-type semiconductors, including germanium, silicon, and indium arsenide. Typical outputs are still just a few millivolts/kilogauss at rated I_c. Although a larger I_c or a smaller d should increase V, these would cause excessive self-heating of the device (by increasing its resistance) and would change its characteristics as well as lower its sensitivity. The resistance of typical Hall devices varies from a few ohms to hundreds of ohms.

SQUIDs *SQUID* stands for superconducting quantum interference device, a superconducting transducer based on the Josephson junction. A SQUID is a thin-film device operating at liquid helium temperature (~4 K), usually made from lead or niobium. The advent of higher-temperature super-

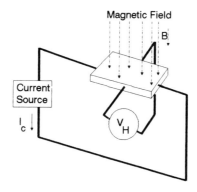

Figure 2-5 Hall effect magnetic field sensor.

conductors that operate in the liquid nitrogen region (~78 K) may produce more practical and inexpensive SQUIDs.

A SQUID element is a Josephson junction that is based on quantum mechanical tunneling between two superconductors. The device is superconducting, with zero resistance, until an applied magnetic field switches it into a normal conducting state, with some resistance. If an external current is applied to the device (and it must be low enough to prevent the current from switching it to a normal conductive state—another Josephson junction property), the voltage across the SQUID element switches between zero and a small value. The resistance and measured voltage go up by steps (or quanta) as the applied magnetic field increases. It measures very small, discrete (quantum) changes in magnetic field strength.

Practical SQUIDs are composed of arrays of these individual junctions and are extremely sensitive magnetometers. For example, they are used to measure small variations in the earth's magnetic field, or even magnetic fields generated inside a living brain.

2.5 Ionizing Radiation Sensors

Ionizing radiation can be particles produced by radioactive decay, such as alpha or beta radiation, or high-energy electromagnetic radiation, including gamma and x-rays. In many of these detectors, a radiation particle (a photon) collides with an active surface material and produces charged particles, ions and electrons, which are then collected and counted as pulses (or events) per second or measured as an average current.

2.5.1 Geiger Counters

When the electric field strength (or voltage) is high enough in a gas-filled tube, electrons produced by primary ionization gain enough energy between collisions to produce secondary ionization and act as charge multipliers. In a *Geiger–Muller tube* the probability of this secondary ionization approaches unity, producing an avalanche effect. So, a very large current pulse is caused by one or very few ionizing particles. The Geiger–Muller tube is made of metal and filled with low-pressure gas (at about 0.1 atm) with a fine, electrically isolated wire running through its center, as shown in Figure 2-6.

A Geiger counter requires a recovery time (dead time) of ~200 microseconds before it can produce another discharge (to allow the ionized particles to neutralize). This limits its counting rate to less than a few kilohertz.

2.5.2 Semiconductor Radiation Detectors

Some *p–n* junction devices (typically diodes), when properly biased, can act as solid-state analogs of an ion chamber, where a high DC voltage across a gas-filled chamber produces a current proportional to the number of ionizing particles striking it per unit time, due to primary ionization. When struck by radiation the devices produce charge carriers (electrons and holes) as opposed to ionized particles. The more sensitive (and use-

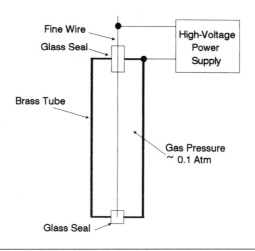

Figure 2-6 Typical Geiger–Muller tube.

ful) devices must be cooled to low temperatures (usually 78 K, by liquid nitrogen).

2.5.3 Scintillation Counters

This device consists of a fluorescent material that emits light when struck by a charged particle or radiation, similar to the action of a photocathode in a photodiode. The emitted light is then detected by an optical sensor, such as a PMT.

2.6 Position (Displacement) Sensors

A wide variety of transducers are used to measure mechanical displacement of the position of an object. Some require contact with the measured object, others do not.

2.6.1 Potentiometers

The *potentiometer* (variable resistor) is often mechanically coupled for displacement measurements. It can be driven by either AC or DC signals and does not usually require an amplifier. It is inexpensive but cannot usually be used in high-speed applications. It has limited accuracy, repeatability, and lifetime, due to mechanical wear of the active resistive material. These devices can either be conventional rotary potentiometers or have a linear configuration with a slide mechanism.

2.6.2 Capacitive and Inductive Sensors

Simple *capacitive* and *inductive* sensors produce a change in reactance (capacitance or inductance) with varying distance between the sensor and the measured object. They require AC signals and conditioning circuitry and have limited dynamic range and linearity. They are typically used over short distances as proximity sensors, to determine if an object is present or not. They do not require contact with the measured object.

2.6.3 LVDTs

The LVDT (*linear voltage differential transformer*) is a versatile device used to measure displacement. It is an inductor consisting of three coils around a movable core, connected to a shaft, as shown in Figure 2-7. The center coil is the transformer's primary winding. The two outer coils are

20 CHAPTER 2 Analog Signal Transducers

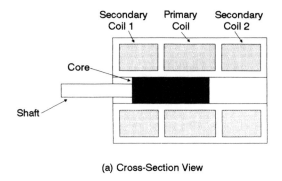

(a) Cross-Section View

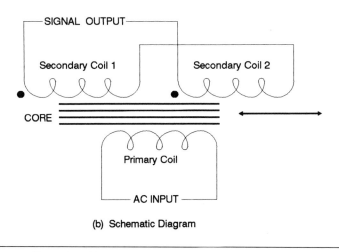

(b) Schematic Diagram

Figure 2-7 Linear variable differential transformer (LVDT).

connected in series to produce the secondary winding. The primary is driven by an AC voltage, typically between 60 Hz and several kilohertz. At the null point (zero displacement), the core is exactly centered under the coils, and the secondary output voltage is zero. If the shaft moves, and the core along with it, the output voltage increases linearly with displacement as the inductive coupling to the secondary coils becomes unbalanced. A movement to one side of the null produces a 0° phase shift between output and input signal. A movement to the other side of null produces a 180° phase shift.

If the displacement is kept within a specified range, the output voltage varies linearly with displacement. The main disadvantages to using an

LVDT are its size, its complex control circuitry, and its relatively high cost.

2.6.4 Optical Encoders

The *optical encoder* is a transducer commonly used for measuring rotational motion. It consists of a shaft connected to a circular disk, containing one or more tracks of alternating transparent and opaque areas. A light source and an optical sensor are mounted on opposite sides of each track. As the shaft rotates, the light sensor emits a series of pulses as the light source is interrupted by the pattern on the disk. This output signal can be directly compatible with digital circuitry. The number of output pulses per rotation of the disk is a known quantity, so the number of output pulses per second can be directly converted to the rotational speed (or rotations per second) of the shaft. Encoders are commonly used in motor-speed control applications. Figure 2-8 shows a simple, one-track encoder wheel.

An *incremental optical encoder* has two tracks, 90° out of phase with each other, producing two outputs. The relative phase between the two channels indicates whether the encoder is rotating clockwise or counterclockwise. Often there is a third track that produces a single index pulse, to indicate an absolute position reference. Otherwise, an incremental encoder only produces *relative* position information. The interface circuitry or computer must keep track of the absolute position.

An *absolute optical encoder* has several tracks, with different patterns on each, to produce a binary code output that is unique for each encoded position. There is a track for each output bit, so an 8-bit absolute encoder has eight tracks, eight outputs, and 256 output combinations, for a resolution of $360/256 = 1.4°$. The encoding is not always a simple binary

Figure 2-8 Simple one-track optical encoder wheel (24 lines = 15 degrees resolution).

counting pattern, since this would result in adjacent counts where many bits change at once, increasing the likelihood of noise and reading errors. A Gray code is often used, because it produces a pattern where each adjacent count results in only one bit change. An absolute encoder is usually much more expensive than a comparable incremental encoder.

2.6.5 Ultrasonic Range Finder

In Chapter 14, an *ultrasonic range finder* is discussed as a noncontact displacement measurement technique. The time it takes an ultrasonic pulse to reflect from an object is measured, and the distance to the object is calculated from that time delay.

2.7 Humidity Sensors

Relative humidity is the moisture content of the air compared to air completely saturated with moisture and is expressed as a percentage.

2.7.1 Resistive Hygrometer Sensors

There are *resistive hygrometer elements* whose resistance varies with the vapor pressure of water in the surrounding atmosphere. They usually contain a hygroscopic (water-absorbing) salt film, such as lithium chloride, which ionizes in water and is conductive with a measurable resistance. These devices are usable over a limited humidity range and have to be periodically calibrated, as their resistance may vary with time, due to temperature and humidity cycling as well as exposure to contaminating agents.

2.7.2 Capacitive Hygrometer Sensors

There are also *capacitive hygrometer elements* that contain a hygroscopic film whose dielectric constant varies with humidity, producing a change in the device's capacitance. Some of these can be more stable than the resistive elements. The capacitance is usually measured using an AC bridge circuit.

2.8 Fluid Flow Sensors

Many industrial processes use fluids and need to measure and control their flow in a system. A wide range of transducers and techniques are commonly used to measure fluid flow rates (expressed as volume per unit time passing a point).

2.8.1 Head Meters

A *heat meter* is a common device where a restriction is placed in the flow tube to produce a pressure differential across it. This differential is measured by a pair of pressure sensors and converted to a flow measurement. The pressure transducers can be any type, such as those previously discussed. The restriction devices include the orifice plate, the venturi tube, and the flow nozzle.

2.8.2 Rotational Flowmeters

Rotational flowmeters use a rotating element (such as a turbine) that is turned by the fluid flow. Its rotational rate varies with fluid flow rate. The turbine blades are usually made of a magnetized material so that an external magnetic pickup coil can produce an output-voltage pulse each time a blade passes under it.

2.8.3 Ultrasonic Flowmeters

Ultrasonic flowmeters commonly use a pair of piezoelectric transducers mounted diagonally across the fluid flow path. The transducers act as a transmitter and a receiver (a multiplexed arrangement), measuring the velocity of ultrasonic pulses traveling through the moving fluid. The difference in the ultrasonic frequency between the "upstream" and "downstream" measurements is a function of the flow rate due to the Doppler effect.

This survey of common transducers and sensors suitable for a data acquisition system is hardly exhaustive. It should give you a feel for the types of devices and techniques applied to various applications and help you determine the proper transducer to use for your own system.

CHAPTER

Analog Signal Conditioning

Nearly all transducer signals must be conditioned by analog circuitry before they can be digitized and used by a computer. This conditioning often includes amplification and filtering, although more complex operations can also be performed on the waveforms.

3.1 Signal Conditioning Techniques

Amplification (or occasionally attenuation) is necessary for the signal's amplitude to fit within a reasonable portion of the ADC's dynamic range. For example, let us assume an ADC has an input range of 0–5 V and an 8-bit output of $2^8 = 256$ steps. Each output step represents $5/256 = 19.5$ mV. If a sensor produces a waveform of 50 mV peak-to-peak (p–p), when directly digitized (by this ADC) it will use only three of the 256 available output steps and be severely distorted. If the sensor signal is first amplified by a factor of 100 (producing a 5-V p–p waveform), it will use the ADC's full dynamic range and a minimum of information is lost. Of course, if it is amplified too much, some of the signal will be clipped and severely distorted, now in a different way.

Filtering must usually be performed on analog signals for several reasons. Sometimes noise or unwanted signal artifacts can be eliminated by filtering out certain portions of the signal's spectra. For example, a system with high gain levels may need a 60-Hz notch filter to remove noise produced by AC power lines. A low-frequency drift on a signal without useful DC information can be removed using a high-pass filter.

Most often, low-pass filters are employed to limit the high end of a waveform's frequency response just prior to digitization, to prevent aliasing problems (which will be discussed in Chapter 4).

Additional analog signal processing functions include modulation, demodulation, and other nonlinear operations.

3.2 Analog Circuit Components

The simplest analog circuit elements are passive components: resistors, capacitors, and inductors. They can be used as attenuators and filters. For example, a simple RC circuit can be used as a high-pass or low-pass filter, as shown in Figure 3-1.

Discrete semiconductor devices, such as diodes and transistors, are commonly used in analog signal-conditioning circuits. Diodes are useful, among other things, as rectifiers/detectors, switches, clamps, and mixers. Transistors are often used as amplifiers, switches, oscillators, phase shifters, filters, and many other applications.

3.2.1 The Operational Amplifier

The most common analog circuit semiconductor component is the *operational amplifier*, called the op amp. This circuit element is usually a monolithic device (an integrated circuit), although hybrid modules, based on discrete transistors, are still used in special applications. The op amp is used in both linear and nonlinear applications involving amplification and signal conditioning.

An op amp, shown in Figure 3-2, consists of a differential voltage amplifier that can operate at frequencies from zero up to several megahertz. It has two inputs, called noninverting (+) and inverting (−), and responds to the voltage difference between them. The part of the output derived from the + source is in phase with the input, while the part from

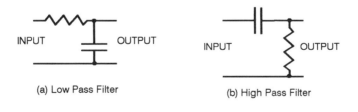

Figure 3-1 Simple RC filters.

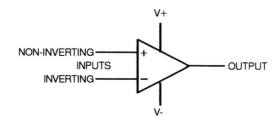

Figure 3-2 The operational amplifier (op amp).

the − source is 180° out of phase. If a signal is applied to both inputs, the output will be zero. This property is called common-mode rejection. Since an op amp can have very high gain at low frequencies (100,000 is typical), a high common-mode rejection ratio prevents amplification of unwanted noise, such as the ubiquitous 60-Hz power-line frequency. Most op amps are powered by dual, symmetrical supply voltages, $+V$ and $-V$ relative to ground, where V is typically in the range of 5 to 15 volts. Some units are designed to work from single-ended supplies. Op amps have very high input impedance (typically a million ohms or more) and low output impedance (in the range of 1 to 100 ohms). An op amp's gain decreases with signal frequency, as shown in Figure 3-3. The point on the gain-versus-frequency curve where its gain reaches 1 is called its unity-

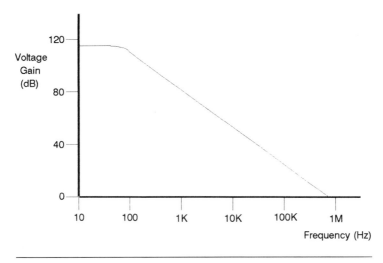

Figure 3-3 Typical op amp gain-versus-frequency curve.

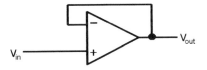

Figure 3-4 Op amp voltage follower.

gain frequency, which is equal to its gain bandwidth product, a constant above low frequencies.

The op amp is more than a differential amplifier, however. Its real beauty lies in how readily its functionality can be changed by modifying the components in its external circuit. By changing the elements in the feedback loop (connected between the output and one or both inputs), the entire characteristics of the circuit are changed, both quantitatively and qualitatively. The op amp acts like a servo loop, always trying to adjust its output so that the difference between its two inputs is zero.

We will examine some common op amp applications here. The reader should refer to the bibliography for other books that treat op amp theory and practice in greater depth.

The simplest op amp circuit is the *voltage follower*, shown in Figure 3-4. It is characterized by full feedback from the output to the inverting input ($-$), where the output is in phase with the noninverting ($+$) input. It is a buffer with very high input impedance and low output impedance.

The *inverting amplifier* shown in Figure 3-5 uses feedback resistor R_2 with input resistor R_1 to produce a voltage gain of R_2/R_1 with the output signal being the inverse of the input. Resistor R_3 is used for DC balance.

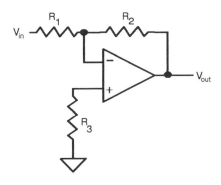

Figure 3-5 Op amp inverting amplifier.

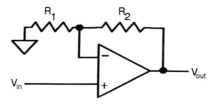

Figure 3-6 Op amp noninverting amplifier.

The *noninverting amplifier* shown in Figure 3-6 uses feedback resistor R_2 with grounded resistor R_1 to produce a voltage gain of $(R_1 + R_2)/R_1$ with the output following the shape of the input (hence, noninverting). Unlike the inverting amplifier, which can have an arbitrarily small gain well below 1, the noninverting amplifier has a minimum gain of 1 (when $R_2 = 0$).

The *difference amplifier* shown in Figure 3-7 produces an output proportional to the difference between the two input signals. If $R_1 = R_2$ and $R_3 = R_4$ then the output voltage is $(V_{in2} - V_{in1}) \times (R_3/R_1)$.

In the *integrator* shown in Figure 3-8, the feedback element is a capacitor (C), producing a nonlinear response. Resistor R_1 and capacitor C have a time constant $R_1 C$. The change in output voltage with time $(dV_{out}/dt) = -V_{in}/(R_1 C)$. Put another way, the output voltage is the integral of $-V_{in}/(R_1 C)dt$. So, this circuit integrates the input waveform. For example, a square-wave input will produce a triangle-wave output as long as the integrator's time constant is close to the period of the input waveform.

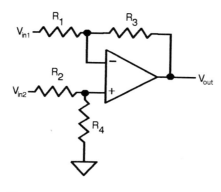

Figure 3-7 Op amp difference amplifier.

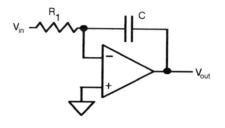

Figure 3-8 Op amp integrator.

Similarly, Figure 3-9 shows a *differentiator*, where the positions of the resistor and capacitor are reversed from those in the integrator circuit. Here, the output voltage = $R_1 C(dV_{in}/dt)$.

More complex op amp circuits include oscillators (both fixed-frequency and voltage-controlled oscillators, or VCOs), analog multipliers and dividers (used in analog computers and modulation circuits), active filters, precision diodes, peak detectors, and log generators.

Many other analog integrated circuits are used as common building blocks in signal-conditioning systems. These ICs include voltage comparators, phase-locked loops, and function generators.

3.2.2 The Voltage Comparator

A *voltage comparator*, as shown in Figure 3-10, is very similar to an op amp used in its highest gain, open-loop configuration (no feedback). Here, if the − input (V_{in}) is greater than the + input (V_{ref}) by at least a few millivolts, the output voltage swings to one extreme ($-V$); if the + input is greater than the − input, the output swings to the other extreme ($+V$). By setting the + or − input to a known reference voltage, an unknown voltage (at the other input) can be evaluated. The comparator can be used

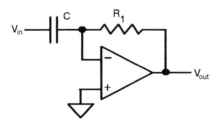

Figure 3-9 Op amp differentiator.

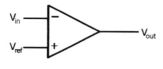

Figure 3-10 Voltage comparator.

to see if analog voltages are within a certain range. It can also be used as a 1-bit ADC.

3.2.3 The Phase-Locked Loop

The *phase-locked loop* is an interesting device. As shown in Figure 3-11, it consists of a phase detector, VCO, and low-pass filter. This comprises a servo loop, where the VCO is phase locked to the input signal and oscillates at the same frequency. If there is a phase or frequency difference between the two sources, the phase detector produces an output that is used to correct the VCO. The low-pass filter is used to remove unwanted high-frequency components from the phase detector's output. One application for this device is to demodulate an FM (frequency modulated) signal.

3.2.4 The Tone Decoder

The *tone decoder* is similar to the phase-locked loop (see Figure 3-12) except that the filtered phase-detector output goes to a comparator instead of feeding back to the VCO. The VCO frequency is constant, so the comparator is activated only when the input signal is within the pass band centered on the VCO frequency. This device is commonly used for frequency detection, as in telephone touch-tone equipment.

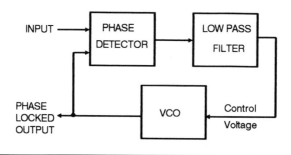

Figure 3-11 Phase-locked loop.

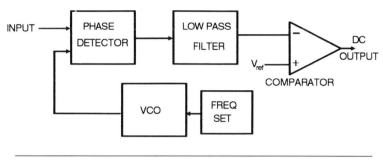

Figure 3-12 Tone Decoder.

3.2.5 The Function Generator

Function generator ICs are special purpose oscillators used to produce sine, square, and triangle waveforms. The signal frequencies are varied either by external resistors and capacitors or by a control voltage, as with a VCO. The output can be frequency modulated by a signal on the VCO input. Some devices also provide for amplitude modulation. These devices can typically produce outputs within the range of 0.01 Hz to 1 MHz. They are often used in test equipment.

Other common analog ICs include a wide range of amplifiers, signal generators, timers, and filters.

3.3 Analog Conditioning Circuits

Analog signal-conditioning circuitry can range from a ridiculously simple RC filter, using two passive components, to a complex system using hundreds of ICs and discrete devices.

3.3.1 Filters

Filtering is undoubtedly the most commonly used analog signal-conditioning function. Usually only a portion of a signal's frequency spectrum contains valid data and the rest is noise. A common example is 60-Hz AC power-line noise, present in most lab environments. A high-gain amplifier will easily amplify this low-frequency noise, unless it is rejected using a band-reject filter or high-pass filter. The standard types of filters are low pass, high pass, band pass, and band reject (or notch filter). The low-pass filter attenuates signals *above* its cutoff frequency, and the high-pass filter attenuates signals *below* its cutoff frequency. The band-pass filter attenuates frequencies outside of its pass-band range (both above and below), and the band-reject filter attenuates those frequencies within its pass-band range. See Figure 3-13 for amplitude-versus-frequency curves of ideal filters.

The study of filters is an entire discipline unto itself. We will only touch on some simple examples here. The reader is referred to the bibliography for more details on the design and use of filters. The two general classes of filters are *active* and *passive*, depending on the components used. A passive filter, using only resistors, capacitors, and inductors, has a maximum gain (or transfer function value) of 1; an active filter, which uses passive components along with active components (usually op

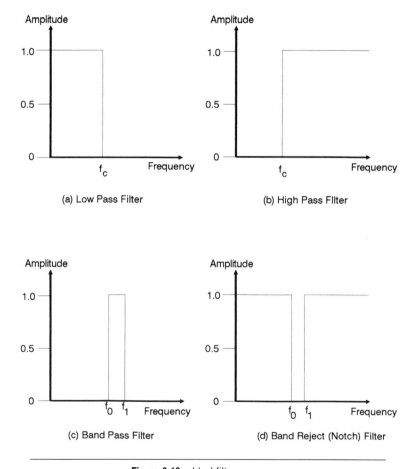

Figure 3-13 Ideal filter responses.

Passive Filters

The simplest filters use a single resistor and capacitor, so they are called *RC* filters. They rely on the frequency-dependent reactance of capacitors for filtering effects. RC circuits are usually used as simple low-pass and high-pass filters. The reactance of an ideal capacitor is $-j/\omega C$ (where $\omega = 2\pi f$, C is capacitance, and $j = \sqrt{-1}$).

The RC low-pass filter is shown in Figure 3-1a. V_{in} is input AC voltage and V_{out} is output AC voltage. The transfer function that describes the response of the circuit is $H(f) = V_{out}/V_{in}$. Since the two components are in series, the current through them is the same: $I_R = I_C$. Z is the AC impedance. Since $V = I \times Z$,

$$H(f) = \frac{(I \times Z_C)}{[I \times (Z_R + Z_C)]}$$

$$= \frac{Z_C}{(Z_R + Z_C)}$$

Since $Z_R = R$ and $Z_C = -j/\omega C$,

$$H(f) = \frac{1}{(1 + j\omega RC)}$$

Note that as frequency (or $\omega = 2\pi f$) approaches zero, the magnitude of the transfer function $|H(f)|$ approaches 1, or no attenuation. Also, the phase angle of $H(f)$ (the phase shift between output and input) approaches zero degrees. As f increases, $|H(f)|$ decreases and the phase angle becomes more negative. The *cutoff frequency* f_c is where the magnitude of the real and imaginary impedance components are equal (when $\omega RC = 1$) and $|H(f)| = 1/\sqrt{2} = 0.707$. This is the -3 dB point [$20 \times \log(0.707) = -3$ dB]. The phase angle at f_c is $-45°$. Well above f_c (i.e., $f > 10 \times f_c$), $|H(f)|$ falls off at -20 dB per decade of frequency (for every frequency increase of $10x$ the voltage output drops $10x$). This is the same as dropping 6 dB per octave (whenever the frequency doubles). At these higher frequencies, the phase shift approaches $-90°$. Now the low-pass filter acts as an *integrator*. It is important to remember that this integration is only accurate at high frequencies (well above cutoff).

The RC high-pass filter, shown in Figure 3-1b, is similar to the low-pass filter just discussed. Here, the output voltage is across the resistor instead of the capacitor. The transfer function for this circuit is $H(f) = 1/[1 - j/(\omega RC)]$. Now, as the frequency gets higher, $|H(f)|$ approaches 1. As the frequency approaches zero, $|H(f)|$ becomes very small.

Again, the 3-dB cutoff frequency f_c is where $\omega RC = 1$. The phase angle at f_c is now $+45°$. At higher frequencies, the phase angle decreases toward 0. At lower frequencies ($f < f_c/10$), the phase angle approaches $+90°$ and $|H(f)|$ increases at the rate of 20 dB per decade. In this low-frequency, high-attenuation region, the RC high-pass filter performs as a *differentiator*. Similar to the RC integrator, this differentiation is only accurate at relatively low frequencies.

Another important point about passive RC integrators and differentiators is that their operational frequency range is in a high-attenuation region. So, their output signals will be very low amplitude, possibly limiting their usefulness because of excessive noise.

RL circuits can also be used as low-pass and high-pass filters, yet they are much less common. A series RLC circuit, as shown in Figure 3-14, is common as a band-pass filter, however. Here, the maximum value of $|H(f)| = 1$ occurs at $f_0 = 1/[2\pi\sqrt{(LC)}]$, where the phase angle is zero. This is the filter's *resonant frequency*. Below f_0, $|H(f)|$ decreases while the phase angle increases toward $+90°$ (as f approaches zero). Above f_0, $|H(f)|$ again decreases, while the phase angle approaches $-90°$. Well above or below f_0, $|H(f)|$ falls off at -20 dB per decade. However, close to f_0 this fall-off may be much steeper, depending on the value of Q, a measure of the filter's resistive losses. $Q = 2\pi f_0 * L/R$. The smaller the value of R is, the larger Q becomes and the steeper the $|H(f)|$ curve becomes—around f_0.

Similarly, a parallel RLC circuit, as shown in Figure 3-15, acts as a band-reject (notch) filter, with a maximum $|H(f)|$ at resonance. This is sometimes referred to as a tank circuit because, at the resonant frequency, it effectively stores most of the electrical energy available (except for losses through the resistor).

Using passive components, if a broader pass-band response or a steeper attenuation curve for out-of-band frequencies is desired, usually several simple filter stages are concatenated. This can produce the desired frequency response at the expense of higher attenuation within the pass

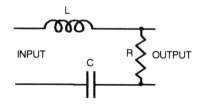

Figure 3-14   Series RLC filter.

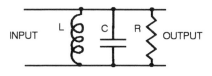

Figure 3-15 Parallel RLC filter.

band, referred to as the insertion loss. One way around this problem is to use an active filter.

Active Filters   Active filters are typically op amp circuits using resistors and capacitors to produce the required frequency response, usually with a gain greater than 1 (no inductors are needed). They are limited to relatively low frequencies (i.e., <1 MHz) due to the limited frequency response of op amps. However, in the audio and ultrasonic regions they are indispensable. Figure 3-16 shows simple active low-pass and high-pass filters.

A newer type of active filter device is the *switched capacitor filter*. This device is very attractive because external component values are not critical (as they are with op amp active filters), and the filter can be tuned by varying the frequency of the applied clock signal (usually a digital waveform). This is a better approach when a computer-controlled filter is required.

3.3.2   Wheatstone Bridge

Many other types of analog circuits are used for conditioning transducer signals. For resistive sensors such as strain gages and thermistors, the classic *Wheatstone bridge* is still used. A DC Wheatstone bridge is shown in Figure 3-17. If the resistance values are set so that there is no voltage across the meter (and no current through it), the bridge is said to be balanced. At balance, it can be shown that $R_1/R_3 = R_2/R_4$. Typically a resistive sensor is placed in a bridge circuit to produce a voltage signal output. Usually, one of the resistors in the bridge is the variable sensor element, and initially the bridge is not balanced. Let us assume for the moment that R_1 is the variable resistive transducer and that for simplicity $R_3 = R_4$. When $R_1 = R_2$ the bridge is balanced and the output is zero. As R_1 increases or decreases slightly, the output voltage will swing positive or negative.

Bridges are also used with AC excitation and reactive elements. This is how a capacitive sensor can produce an accurate voltage signal. In

36 CHAPTER 3 Analog Signal Conditioning

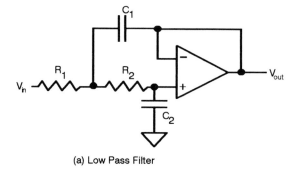

(a) Low Pass Filter

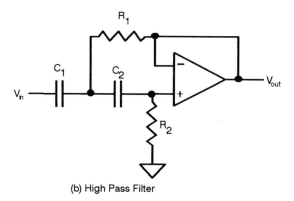

(b) High Pass Filter

Figure 3-16 Active filters based on op amps.

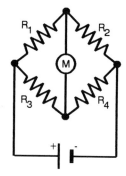

Figure 3-17 Wheatstone bridge.

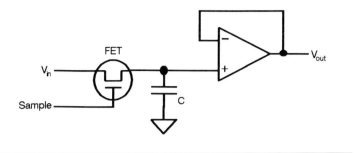

Figure 3-18 Sample-and-hold amplifier.

the case of an AC bridge, usually one leg is left as purely resistive, making it easier to balance the unknown reactive element in the other leg.

3.3.3 The Sample-and-Hold Amplifier

Another special analog circuit, extremely useful in data acquisition applications, is the *sample-and-hold amplifier* shown in Figure 3-18. This is used to get a stable sample of a changing analog signal, prior to using an ADC. The field-effect transistor (FET) acts as a switch, charging the capacitor to the analog signal's present voltage level when the sample line is asserted. When the transistor is switched off, the capacitor "remembers" the voltage, which is buffered by the op amp. The very high input impedance of the op amp, along with a low-leakage capacitor, prevents the voltage from dropping off too quickly.

A sample-and-hold amplifier is used as the front end of an ADC so that if the analog waveform is rapidly changing during the ADC cycle, the value produced can have a large error. This way, there is an accurate "snapshot" of the waveform during the brief sample interval. The sample interval is typically much shorter than the time between successive analog conversions.

3.3.4 Peak Detector

Another useful circuit is the *peak detector*, shown in Figure 3-19, which again is op-amp based. It is similar to the sample-and-hold circuit, with a diode used as a switch for charging the capacitor C_1. The second (output) op amp is simply a buffer, allowing the circuit to drive a low-impedance load without draining the capacitor. Whenever the input voltage is greater than the output voltage, the diode is forward biased and the capacitor is

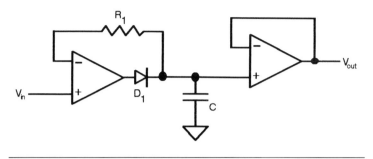

Figure 3-19 Peak detector.

charged up to that voltage. Usually a switch (such as a FET) may be placed across the capacitor to implement a discharge or reset function.

3.3.5 Log and Antilog Amplifiers

There are many important nonlinear amplifier circuits, including the log amplifier and the antilog amplifier. The simple logarithmic amplifier uses a junction diode as a nonlinear element. In a forward-biased diode, the voltage drop across the diode varies proportionally to the log of the current through it. When a diode is connected in the feedback loop of an inverting amplifier, the output voltage is a logarithmic function of the input voltage. If a diode is used in a noninverting amplifier, the result is an antilog amplifier.

There are some problems using diodes in log amplifiers. They are very temperature sensitive, since the forward voltage drop across a diode is a function of temperature. In fact, this property is often exploited in

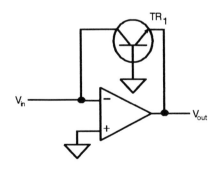

Figure 3-20 Simple logarithmic amplifier.

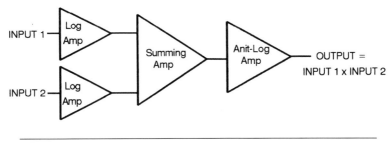

Figure 3-21 Analog multiplier.

diode temperature sensors. Also, the signal range over which the diode has a logarithmic response is somewhat limited. Often a bipolar transistor is used in place of a diode, since its emitter-base voltage varies with the log of its collector current over a very wide range. The log amp circuit using a transistor is shown in Figure 3-20.

3.3.6 Modulation

Another important nonlinear function is modulation. Frequency modulation was discussed with the VCO. Amplitude modulation is easily achieved using an *analog multiplier*. A simple means of producing an analog multiplier is shown in Figure 3-21. The two inputs each pass through a log amplifier and then are added together; finally, they pass through an antilog amplifier. The output voltage is equal to the product of the input voltages times a scaling factor. Analog multipliers are commonly available as single-chip devices.

There are many other standard analog signal conditioning circuits besides the ones shown here. This chapter should give you a feel for what is commonly available and help you locate more detailed information as you require it.

CHAPTER

Analog/Digital Conversions

As previously noted, we live in an analog world. Nearly all "real-world" measured quantities are analog, at least at the macroscopic level we typically deal with. Analog waveforms are usually defined as smooth, continuous functions that have derivatives existing nearly everywhere. Most transducers have analog outputs, usually voltage or current, which represent the physical quantities being measured, such as temperature or pressure (a notable exception is the optical encoder with its digital output). Whenever an analog quantity is discussed here, it refers to a voltage or current suitable for use with common electronic equipment. This is typically in the frequency range of zero to 1 MHz, with a voltage range of around 1 microvolt (μV) to 100 V or a current range of about 1 microampere to 10 amps.

4.1 Digital Quantities

Digital quantities have discrete levels that vary by steps instead of continuously (as shown in Figure 1-1 of Chapter 1). Most digital electronic equipment uses binary values, which have two possible states called true (on or 1) and false (off or 0). Most often the 0/1 notation is used to describe the binary level of a single line or wire, represented as a binary digit or *bit*. For the standard family of TTL (*transistor transistor logic*) digital ICs, a high level (>2.4 V) is a logical 1 and a low level (<0.8 V) is a logical 0.

Binary values are a base-2 numbering system, as opposed to our everyday base-10 decimal system. It takes many bits grouped together to represent a useful quantity. In general, a collection of n bits can represent 2^n discrete levels. For example, a group of eight bits is referred to as a byte, where $2^8 = 256$ levels, for a representation of values in the range of 0 to 255 (or -128 to $+127$). A group of 16 bits is referred to as a word, having $2^{16} = 65,536$ steps. In digital electronic equipment, these groups of bits are usually parallel lines or wires, where each bit is present at the same time. One wire typically carries the value for one bit. This means that increasing the number of levels a digital circuit can represent increases the number of wires (or interconnections) in that circuit. This also allows the digital representation to more closely approximate the analog signal, within a given dynamic range.

The concept of *dynamic range* is very important for data acquisition systems; it will be addressed at greater length in Chapter 10. By definition, the dynamic range of a data acquisition system is the ratio of the maximum value that can be measured to the smallest value that can be resolved. This number is often represented in decibels (dB) as

$$\text{Dynamic range (dB)} = 20 * \log_{10}(\text{max}/\text{min})$$

If both positive and negative values are measured,

Maximum value = maximum positive value − minimum negative value.

For example, a data acquisition system with a 1-millivolt resolution and a value range of 0 to $+10$ volts (or -5 to $+5$ volts) has a dynamic range of $10,000:1$, or 80 dB. This dynamic range requires a minimum of 14 bits to represent it, since $2^{14} = 16,384$, which is greater than 10,000, while 2^{13} (8192) is less than 10000.

4.1.1 Binary Codes

For n binary lines to represent 2^n levels, each line must have a different value or weight. For a *natural binary* code, having any value from 0 to $2^n - 1$, integers are represented by a series of weighting bits having the value 2^m (where m varies from 0 to $n - 1$). The bit number m is zero for the least significant bit (LSB) on the far right and increases to $n - 1$ for the most significant bit (MSB) on the far left. The values of integer bit weights for the first 16 bits are given in Table 4-1. The value of a collection of parallel bits is the sum of the weighted values of all nonzero bits (or the value of a bit, either 0 or 1, times its weight). For example, we will evaluate the 8-bit binary integer 01011101. Starting with the LSB, working from right to left:

TABLE 4-1
Positive Integer Bit Weights for Natural Binary Code

BIT # (m)	BIT WEIGHT (2^m)
0	1
1	2
2	4
3	8
4	16
5	32
6	64
7	128
8	256
9	512
10	1024
11	2048
12	4096
13	8192
14	16384
15	32768

$$\text{Sum} = 1*2^0 + 0*2^1 + 1*2^2 + 1*2^3 + 1*2^4 + 0*2^5 + 1*2^6 + 0*2^7$$
$$= 1 + 0 + 4 + 8 + 16 + 0 + 64 + 0$$
$$= 93$$

Sometimes it is necessary to represent both positive and negative integer values, as when dealing with a bipolar voltage. The most common binary code for this is called twos complement, which can represent values from -2^{n-1} to $+2^{n-1} - 1$. In this notation, positive values are encoded the same way as the positive-only, natural binary code, above (this includes zero). To encode a negative value, write down the code for the corresponding positive value (including all leading zeros), invert the number by changing all ones to zeros and all zeros to ones (which is called the ones complement), and then add one to the result. Table 4-2 contains twos complement codes for 5-bit numbers representing values $+15$ to -16. For example, to get the twos complement representation of the value -12 using five bits:

1. $+12 = 01100$
2. Ones complement $= 10011$
3. Twos complement $= 10011 + 1$
4. $-12 = 10100$

One additional coding system we will mention here is *fractional binary*. This is useful when digital readings must be normalized to an arbitrary full-scale value, as when a converter's reference voltage is variable. The n bits of the code represent values between 0 and $1 - 2^{-n}$. The weight of each bit is a fractional value, equal to its natural binary integer value (of 2^m) divided by 2^n. This means the MSB has a weight of $\frac{1}{2}$ (since $2^{n-1}/2^n = 2^{-1}$), the next bit to the right has a weight of 1/4, and so on, down to the LSB with a weight of $1/2^n$ (or 2^{-n}). When all bit values are 1, the total value is $1 - 2^{-n}$. Again, 2^n levels are represented by this code. Table 4-3 lists fractional binary codes for 5-bit values. Note that sometimes fractional binary values are written with a *binary point* and sometimes not. So, the fractional binary for 1/32 can be written as either 0.00001 or 00001, even though they both mean the same thing.

TABLE 4-2
Twos Complement Coding for 5-Bit Bipolar Values

VALUE	TWOS COMPLEMENT CODE
+15	01111
+14	01110
+13	01101
+12	01100
+11	01011
+10	01010
+9	01001
+8	01000
+7	00111
+6	00110
+5	00101
+4	00100
+3	00011
+2	00010
+1	00001
0	00000
-1	11111
-2	11110
-3	11101
-4	11100
-5	11011
-6	11010
-7	11001
-8	11000
-9	10111
-10	10110
-11	10101
-12	10100
-13	10011
-14	10010
-15	10001
-16	10000

TABLE 4-3
Five-Bit Fractional Binary Codes

CODE	FRACTION OF FULL SCALE
0.00000	0
0.00001	1/32 (LSB)
0.00010	2/32 = 1/16
0.00011	3/32
0.00100	4/32 = 1/8
0.00101	5/32
0.00110	6/32 = 3/16
0.00111	7/32
0.01000	8/32 = 1/4
0.01001	9/32
0.01010	10/32 = 5/16
0.01011	11/32
0.01100	12/32 = 3/8
0.01101	13/32
0.01110	14/32 = 7/16
0.01111	15/32
0.10000	16/32 = 1/2 (MSB)
0.10001	17/32
0.10010	18/32 = 9/16
0.10011	19/32
0.10100	20/32 = 5/8
0.10101	21/32
0.10110	22/32 = 11/16
0.10111	23/32
0.11000	24/32 = 3/4
0.11001	25/32
0.11010	26/32 = 13/16
0.11011	27/32
0.11100	28/32 = 7/8
0.11101	29/32
0.11110	30/32 = 15/16
0.11111	31/32

4.2 Data Conversion and DACs

Data conversion is at the heart of data acquisition systems. Real-world analog signals must be converted to binary representations via an *analog-to-digital converter* (or ADC). Similarly, if output to the analog world is required, as in control systems, digital values are transformed using a digital-to-analog converter (or DAC). We will look at DACs first, because they are usually simpler devices than ADCs. In addition, many ADCs contain DACs as part of their circuitry.

DACs use either current or voltage switching techniques to produce an output analog value equal to the sum of several discrete analog values. Because it is easier to sum currents (rather than voltages) using analog circuitry, most commonly available DACs are current-mode devices. They produce the sum of internal current sources and use either an internal or external op amp as a current-to-voltage converter.

4.2.1 Fully Decoded DAC

One type of DAC is shown in Figure 4-1. This is a fully decoded current-mode 3-bit DAC. A fully decoded DAC, for n input bits, contains $2^n - 1$ switches and identical current sources. Basically, the input bits are decoded and control switches to the current sources of equal magnitude. A digital value of 001 connects one current source to the output, a value of 010 connects two sources to the output, 011 connects three sources to the output, and so on up to seven sources for 111. These current sources are summed at the output, producing a current proportional to the digital value.

The main advantage to this type of fully decoded DAC is that with proper switching the output current is guaranteed to be *monotonic*. That is, as the digital code continues to increase the analog output will also increase, step by step. This is not always true of all DACs. The disadvantage to this type of DAC is that $2^n - 1$ current sources and switches are required. This becomes prohibitive for reasonably large numbers of bits, such as 4095 current sources for a 12-bit DAC.

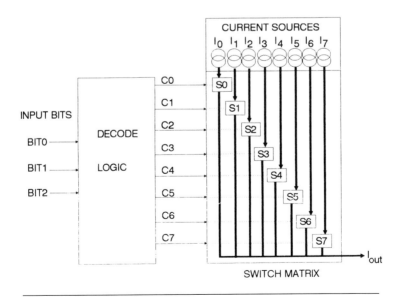

Figure 4-1 Fully decoded 3-bit current mode DAC.

46 CHAPTER 4 Analog/Digital Conversions

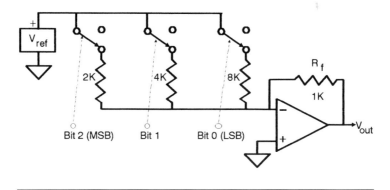

Figure 4-2 Weighted resistor, 3-bit current mode DAC.

4.2.2 Weighted Resistor DAC

A simpler DAC can be produced using a voltage reference with a set of weighted precision resistors and switches, as shown in the 3-bit DAC example in Figure 4-2. The resistor values are in a binary bit-weight ratio (1:2:4:8:16 and so on). Again, this converter is a current-mode device, with the sum of all resistor currents resulting in an analog current.

In this example, as in nearly all practical current-mode DACs, the output current is passed through an op amp. This acts as a current-to-voltage converter as well as isolating the DAC from output circuit loading. Here, since the op amp is inverting (because the virtual ground of the inverting input is needed), the output is a negative voltage proportional to the input binary word and the voltage reference.

When all input bits are zero, no current flows into the op amp, and the output voltage is zero. If the MSB (bit 2) is 1, the current flowing into the op amp is $V_{ref}/2K$, producing an output voltage of $-V_{ref}/2$, since the feedback resistor (R_f) is $1K$ ohm and the op amp's gain is $-R_f/R_{in}$. Similarly, if bit 1 is 1, it feeds a current of $V_{ref}/4K$, producing an output voltage of $-V_{ref}/4$; and if the LSB (bit 0) is 1, it feeds a current of $V_{ref}/8K$, producing an output voltage of $-V_{ref}/8$. If more than a single bit is 1, their currents sum at the op amp's input and produce the appropriate output voltage. If all bits are 1, the output voltage is $-7/8\ V_{ref}$. This is the full-scale output.

This DAC can produce eight discrete analog output levels, spaced $\frac{1}{8}V_{ref}$ apart. Note that if we treat these values as normalized to V_{ref}, we are dealing with fractional binary values. If we set $V_{ref} = 10.00$ V, the full-scale output is -8.75 V, with steps of 1.25 V. If we increased the number of bits in this DAC to n, the resistor values for the most significant bits would stay the same, and larger resistors would be added for the least significant bits. The LSB would have a value of $2^n * 1K$ ohm.

The advantage of the DAC in Figure 4-2 is that only one switch and resistor are needed per bit. The main drawbacks are that as the number of converter bits increases, the number of different precision resistor values needed, as well as the overall range of resistor values, increases. If we increased the resolution of the DAC in Figure 4-2 from three bits to eight bits, the resistance values would increase up to $256K$ ohms. This makes it very difficult to maintain monotonicity, linearity, and overall accuracy, due to the wide range of resistance values required.

4.2.3 Resistor Quad

Other techniques are used to overcome these drawbacks. One of these is the *binary resistance quad*, used in an 8-bit DAC in Figure 4-3. Here, the resistor network uses the same four values for more than four bits resolution. The resistors and switches constitute a voltage-divider network. The most significant four bits (bits 4–7) are in the usual scaled binary ratio of $2:4:8:16$. The least significant four bits (bits 0–3) are these same values, repeated. However, these values are attenuated $16:1$, via the additional ($16K$-ohm) resistor. Each section of four resistors is called a quad.

4.2.4 R–2R Ladder

A very common DAC uses the R–$2R$ resistance ladder, where only two different resistor values are needed, as shown in Figure 4-4. When only the MSB (bit 7) is 1, the output voltage is $-V_{\text{ref}}/2$, since V_{ref} is switched

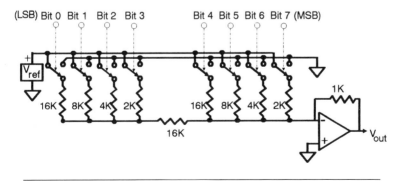

Figure 4-3 Eight-bit DAC using resistor quads.

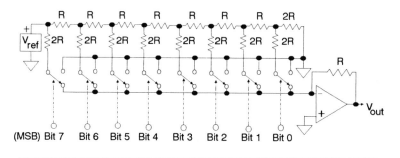

Figure 4-4 Eight-bit DAC using R-2R resistor ladder.

through $2R$ from bit 7 and the op amp's feedback resistor is R. When moving down the ladder (toward less significant bits), each $2R$ resistor sees one-half the voltage of the one above it (when it is the only 1 bit). This is due to the constant resistance of the attenuator network to ground. So, bit 6 contributes $-V_{ref}/4$ to the output voltage, bit 5 contributes $-V_{ref}/8$, and so on down to bit 0 contributing $-V_{ref}/256$.

4.2.5 Multiplying DAC

When a DAC can operate with a variable analog reference voltage instead of the usual fixed value, it is called a multiplying DAC. The output of this DAC is proportional to both the analog reference input and the digital input. If it can respond to bipolar inputs (both analog and digital) and produce a bipolar output, it is a four-quadrant multiplying DAC. This refers to a Cartesian plot of the transfer function. A multiplying DAC is commonly used as a digitally controlled attenuator or amplifier of an analog signal.

4.2.6 DAC Characteristics

Some important criteria must be considered when choosing a DAC. The first parameter to determine is the number of bits of resolution. This is selected by knowing the desired dynamic range of the output signal. Eight and 12-bit DACs are commonly available as *monolithic devices* or integrated circuits (ICs).

Another major parameter is *settling time*, which determines the speed of conversion, as shown graphically in Figure 4-5a. This is the amount of time required for a DAC to move to and stay within its new output value (usually to $\pm\frac{1}{2}$ LSB) when the digital input changes. For current output DACs, settling time is usually quite fast, typically a few hundred nanoseconds. If a fast-settling op amp is used as an output current-to-voltage converter, output waveforms at frequencies well over 1 MHz can be produced.

Linearity is another major DAC parameter. It is the maximum deviation of the DAC's transfer curve from an ideal straight line, usually expressed as a fraction of the full-scale reading, as illustrated in Figure 4-5b.

One final DAC parameter to note is *monotonicity*. If the output of a DAC always increases for increasing digital input, the DAC is considered monotonic. Monotonicity is specified over a certain number of input bits, typically the full number of bits of resolution. A nonmonotonic DAC would have a dip in its transfer curve.

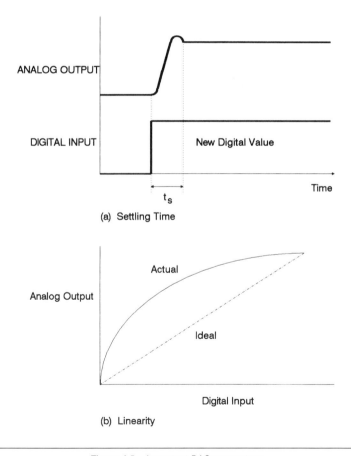

Figure 4-5 Important DAC parameters.

4.3 ADCs

Now we will turn our attention to ADCs. A multitude of techniques are used to produce an analog-to-digital converter. We will look at some of the more common ones here.

4.3.1 Ramp ADC

One of the simpler approaches in implementing an ADC is the *ramp converter* shown in Figure 4-6. It consists of a digital counter, a DAC, an analog comparator, and control logic with timing generation. Basically, when an analog conversion is requested, the digital counter starts counting up from zero. As it counts, the analog output of the DAC increases, or ramps up. When the DAC's output is equal to or exceeds the analog input, the comparator's output switches and the control logic stops the counting. An end of conversion is indicated, with the digital counter output now containing the converted value. This conversion sequence is illustrated in Figure 4-7.

The problem with this technique is its relatively long conversion time, or slow speed, which becomes worse with increasing number of output bits. Everything else being equal, the maximum conversion time for the ramp converter increases as 2^n, where n is the number of bits of resolution. The conversion time is inversely proportional to the frequency of the clock used in counting.

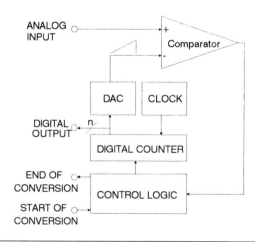

Figure 4-6 Simple ramp analog-to-digital converter (ADC).

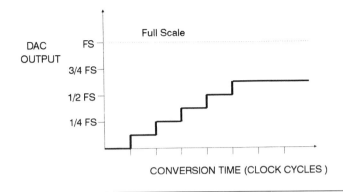

Figure 4-7 Ramp ADC, typical conversion sequence.

For example, if the converter's DAC had a 200-nsec settling time and we used a 5-MHz clock for a 12-bit ADC, maximum conversion time would be $\frac{1}{5 \times 10^6} \times 4096 = 819.2$ μsec. This would allow a conversion rate of only 1220 samples per second. Of course, this is a worst-case value. If the analog input is less than the maximum allowable value, conversion time will be shorter.

One minor variant on this technique is the *servo ADC*. Its digital counter can count both up and down. When the DAC output is below the analog input, it counts up. When the DAC output is above the analog input it counts down. It tends to track the analog input continuously, analogous to a servo control loop. It will respond to small input changes rapidly, but it is as slow as the standard ramp converter when a large input change has occurred.

4.3.2 Successive-Approximation ADC

A major improvement on the ramp converter is the *successive-approximation converter*, probably the most popular class of ADC commercially available at present. The overall block diagram of this system is very similar to the ramp converter's, as shown in Figure 4-8, except that the digital counter is replaced by more sophisticated control logic that includes a shift register. Instead of simply counting up until the analog value is exceeded, the successive-approximation ADC tests one bit at a time (starting with the most significant) until the internal DAC value is as close as possible to the analog input without exceeding it.

First, the most significant bit (MSB), equal to 1/2 full-scale (FS) value, is turned on; if the DAC's output is less than the analog input, it is

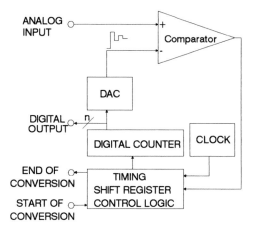

Figure 4-8 Simple successive approximation ADC.

left on (otherwise it is turned off). Then the next bit down (1/4 FS) is turned on and left on only if the DAC's output is still less than the analog input. This process continues until all n bits have been tested. Figure 4-9 shows a typical conversion sequence. The entire conversion requires much less than 2^n clock cycles. Furthermore, the conversion time is relatively constant and insensitive to the input analog value, as opposed to ramp converters.

It is not unusual to find successive approximation ADCs with conversion rates as high as 1 million samples/second and resolution as high as 16 bits. Lower-speed and lower-resolution successive approximation

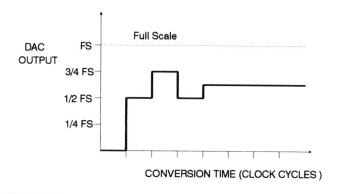

Figure 4-9 Successive approximation ADC, typical conversion sequence.

4.3 ADCs

ADCs are common commercial ICs, available at very low prices. For example, there are 8-bit devices with conversion times of 100 μsec or less (i.e., 10-kHz sampling rates) available for only a few dollars.

4.3.3 Dual-Slope ADC

Another common ADC is the *dual-slope converter*, which relies on integration. As shown in Figure 4-10a and 4-10b, the voltage to be measured (V_x) is input to an integrator, charging the capacitor for a fixed time

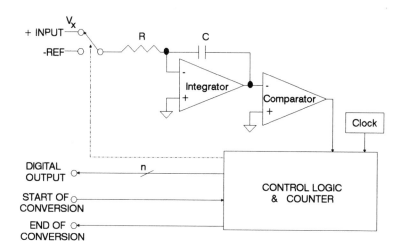

(a) Block Diagram

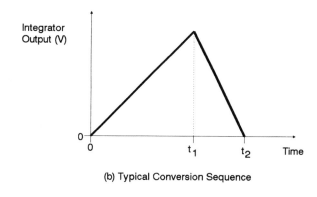

(b) Typical Conversion Sequence

Figure 4-10 Dual-slope ADC.

interval t_1, which corresponds to a certain number of clock cycles. At the end of this interval, a known reference voltage (V_r) of opposite polarity is applied to the integrator, discharging the capacitor. The time (and number of clock cycles) required to bring the integrator output back to zero ($t_2 - t_1$) is measured.

The charge on the capacitor at time t_1 is proportional to the average value of V_x times t_1. This is equal to the charge lost by the capacitor during time $t_2 - t_1$ while being discharged by the reference voltage, proportional to V_r times ($t_2 - t_1$). Hence, ($t_2 - t_1$)/t_1 is proportional to V_x/V_r. The output binary count for the time interval ($t_2 - t_1$) is thus proportional to V_x, the input voltage. With appropriate circuitry, bipolar voltages can also be measured.

The dual-slope ADC has many advantages. Noise present on the input voltage is reduced by averaging. The value of the capacitor and conversion clock do not affect conversion accuracy, since they act equivalently on the up-slope and down-slope. Linearity is very good and extremely high-resolution measurements can be obtained. Its main disadvantage is a slow conversion rate, often in the range of 10 samples/second. In applications where this is not a problem, such as in measuring temperature transducers, a dual-slope ADC is a good choice. They are commonly used in digital voltmeters (DVMs).

4.3.4 Voltage-to-Frequency Converter

Another slow ADC is the *voltage-to-frequency converter*, or VFC. It changes an analog signal into a digital pulse train with a frequency proportional to the signal voltage. This pulse train can be converted into a usable digital output of n parallel bits by clocking a counter for a fixed time interval.

The VFC is an integrating device with good noise rejection and monotonicity, similar to the dual-slope converter. It can also be used as an inexpensive, high-resolution ADC with slow conversion rates. Its drawbacks include nonlinearity, a limited input-voltage dynamic range, and *output offset*. As the input voltage approaches zero, the output frequency is still offset from zero.

4.3.5 Flash ADC

The fastest type of ADC is the *flash converter*. An n-bit flash ADC applies the input voltage to an array of $2^n - 1$ comparators, via a ladder of 2^n resistors. The thresholds for the comparators are spaced 1 LSB apart.

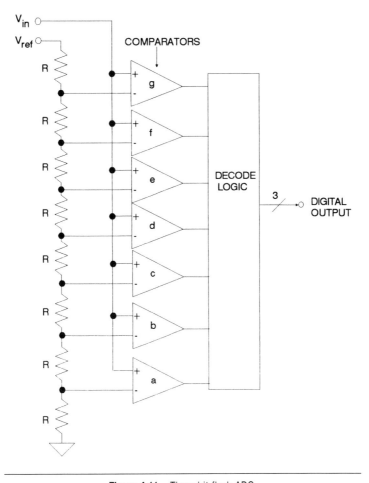

Figure 4-11 Three-bit flash ADC.

Figure 4-11 shows a simple 3-bit flash ADC. When V_{in} is zero, all comparators are off. As the input voltage increases to $V_{ref}/8$, the lowest comparator (a) goes on. As V_{in} keeps increasing by steps of $V_{ref}/8$, each successive comparator (b, c, d, ...) switches on. All comparators are on when the input voltage reaches or exceeds $7/8 * V_{ref}$. The digital logic decodes the comparator outputs into a 3-bit word. The digital output can either be normal binary code (000 = minimum value, 111 = maximum value) or a Gray code. In a Gray code, only one output bit changes for each one-step input change, to minimize noise and ''glitches'' when many digital switches change at once at high speed.

The conversion speed of a flash ADC is limited only by the speed of its comparators and digital logic circuitry. It has a conversion rate measured in speeds of millions of samples per second. A common application for this device is digitizing video signals at rates of 10M samples/second or above. Flash ADCs are very expensive devices when high digital resolution is required, since their complexity grows geometrically with the number of bits ($2^n - 1$ comparators for n bits). So, even an 8-bit flash converter requires 255 comparators and a moderately complex digital decoder.

4.3.6 Sigma-Delta Converter

One of the newest commercial converters is the Sigma-Delta ADC. This device is a low-cost, high-resolution ADC suitable for low conversion rates. A typical Sigma-Delta ADC has 16-bit resolution with an input signal frequency range of 0–10 Hz.

A block diagram of a Sigma-Delta converter appears in Figure 4-12. It consists of an analog modulator loop followed by a digital filter. The modulator operates at a very high clock frequency, effectively oversampling the input signal. It produces a serial data stream, which the digital filter averages to produce a 16-bit output word.

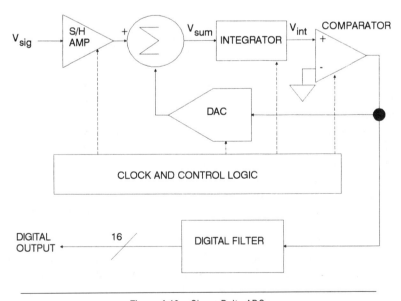

Figure 4-12 Sigma-Delta ADC.

TABLE 4-4
Sigma–Delta Converter, Internal Cycles

CLOCK CYCLE	V_{sum}	V_{int}	COMPARATOR	DAC OUT	
0	+0.4	+0.4	1	+1.0	
1	-0.6	-0.2	0	-1.0	
2	+1.4	+1.2	1	+1.0	
3	-0.6	+0.6	1	+1.0	
4	-0.6	0	1	+1.0	
5	-0.6	-0.6	0	-1.0	Full Conversion Cycle
6	+1.4	+0.8	1	+1.0	
7	-0.6	+0.2	1	+1.0	
8	-0.6	-0.4	0	-1.0	
9	+1.4	+1.0	1	+1.0	
10	-0.6	+0.4	1	+1.0	
11	-0.6	-0.2	0	-1.0	

For example, assume the analog signal range (V_{sig}) is -1.0 V to $+1.0$ V, as well as the DAC output, and the input signal voltage is constant at $+0.4$ V. The comparator's output will be high and the DAC's output will be $+1.0$ V if the output of the integrator (V_{int}) is positive. The comparator's output will be low and the DAC's output will be -1.0 V if V_{int} is negative.

Let us follow the voltages at V_{sum} (where the DAC output is summed with the input signal), V_{int} (the integrator output, where V_{sum} is averaged), and the DAC output, as we step through the first few clock cycles, as shown in Table 4-4. Note that the DAC is a single-bit device, with an output of either $+1.0$ V or -1.0 V.

Initially, at clock cycle 0, we assume that the DAC output is turned off, $V_{sig} = V_{sum} = V_{int}$ ($+0.4$ V), and the comparator output is 1, producing a DAC output of $+1.0$ V, to be subtracted from V_{sum} on the next clock cycle. At clock cycle 1, the first full clock cycle, $V_{sum} = V_{sig} - V_{DAC} = +0.4$ V $- 1.0$ V $= -0.6$ V. V_{int} is simply the previous value of V_{int} plus the new value of V_{sum}, or $+0.4$ V $+ (-0.6$ V$) = -0.2$ V. This process continues until the values at clock cycle 1 occur again, and the process is repeated. In this example, the conversion process starts repeating at clock cycle 11. Hence, 10 clock cycles are required to complete the conversion. If the analog voltage of the DAC output is averaged over those 10 cycles, we get a value of $+4.0/10 = +0.4$ V, the value of V_{sig}. Since the digital filter sees the same numbers as the DAC, its output will also be $+0.4$ V, but as a digital representation.

Note that the number of clock cycles required for conversion varies with the value of V_{sig}. If we used a V_{sig} value of $+0.2$ V, only five clock cycles would be required. So, if high resolution at low sampling rates is adequate, the Sigma-Delta ADC is a good selection and a strong competitor to dual-slope ADCs.

4.3.7 Other ADC Trends

One newly emerging trend to be aware of in data conversion is the sensor-specific ADC. Some converters are being designed to work with particular types of transducers. They already exist for LVDTs and should soon be appearing for other classes of sensors.

4.3.8 ADC Characteristics

After exploring some of the common ADC techniques, a discussion of their major characteristics is in order. The most important ADC parameters are resolution and sampling rate.

ADC Resolution   An ADC's *resolution* is the smallest change it can detect in a measurement. This value is actually a percentage of the full-scale reading, but it is commonly specified as the number of output bits. An n-bit ADC has 2^n possible output values and a resolution of 1 part in 2^n. For example, a 10-bit ADC has a resolution of approximately 0.1% (1/1024). High resolution (more bits) is usually desirable in an ADC. Note that an ADC's accuracy can be no better than its resolution, for an individual reading.

ADC Sampling Rate   *Sampling* or *conversion rate* is the ADC specification most often examined. It is the number of readings completed every second. This parameter is extremely important when rapidly changing signals are measured. It is obvious that if a signal frequency is higher than the sampling rate, rapid signal variations can be missed when they occur between consecutive ADC samples. This is true whether the ADC takes an instantaneous analog measurement, using a sample-and-hold amplifier to keep the value constant for the conversion cycle, or whether the signal value is averaged (with an integrator) during the conversion cycle. In fact, a successive-approximation ADC can produce highly erroneous results if the input signal varies significantly during a conversion cycle.

The Nyquist Theorem For an analog signal to be accurately digitized by an ADC, it must be sampled at a rate at least two times the highest frequency component in that signal. To put it another way, only signals whose highest frequency components are no more than one-half the sampling frequency can be accurately digitized. This maximum signal frequency is called the Nyquist frequency, and this rule is called the Nyquist theorem.

Aliasing When a signal is sampled too slowly (it contains frequency components above the Nyquist frequency), the digitized waveform is distorted. This distortion is called aliasing. It is the result of mixing or beating between the signal frequencies and the sampling frequency. Low-frequency harmonics composed of the differences between the signal and sampling frequencies are recorded instead of the signal itself.

Figure 4-13 shows a simplified example of aliasing, using a single-frequency signal. Figure 4-13a shows a sine wave of fixed frequency f_0. If that signal was digitized at a rate of $2f_0$, the samples take would produce a waveform with a frequency of f_0, as shown in Figure 4-13b. The only distortion here is that the digitized waveform appears to be a triangle wave instead of a sine wave. If a sampling rate much higher than $2f_0$ was used, the digitized waveform would "fill in" more, and it would better approximate a sine wave. If the signal was digitized at a rate of only $(4/3)f_0$, the samples would produce a waveform of frequency $(1/3)f_0$, as shown in Figure 4-13c. This result of aliasing is the difference frequency between the sampling rate and signal frequency, which is $(4/3 - 1) \times f_0$. If the sampling rate was equal to the signal frequency, the digitized waveform would be a constant value.

In general, an ADC's sampling rate should be much higher than twice the maximum signal frequency. A value of five times is a good choice. In most data acquisition systems, the analog input is filtered to eliminate any signal components above the Nyquist frequency. This is often referred to as an anti-aliasing filter. For such a low-pass filter to produce adequate attenuation at the Nyquist frequency, it should have a cutoff frequency well below that point, requiring a sampling rate many times higher than the maximum frequency of interest.

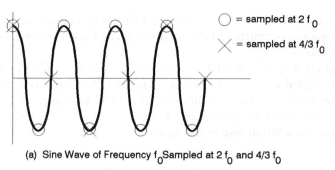

(a) Sine Wave of Frequency f_0 Sampled at $2 f_0$ and $4/3 f_0$

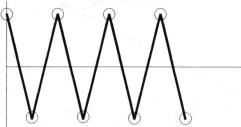

(b) Waveform Reconstructed From $2 f_0$ Samples

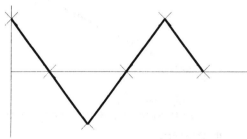

(c) Waveform Reconstructed From $4/3 f_0$ Samples

Figure 4-13   Example of aliasing.

ADC Accuracy   Another important ADC characteristic is its *absolute accuracy*, which is the measure of all error sources. This is sometimes referred to as the total unadjusted error. It is the difference between the ideal input voltage and the actual input voltage (range) to produce a given output code, usually expressed as a percentage of full scale (i.e., ±1 LSB). It is possible for a converter's absolute accuracy to be better than its resolution. By definition, a converter's resolution is 1 LSB. It is not uncommon to find a commercial ADC with an ideal absolute accuracy of ±0.5 LSB. The sources contributing to the total unadjusted error include offset and linearity errors.

An error-free 3-bit ADC transfer curve is displayed in Figure 4-14a, showing digital output code versus analog input voltage as a fraction of full-scale input. As the resolution of the ADC increases, the "coarseness" of this curve decreases and it approaches a straight line, shown as the infinite resolution line in the figure.

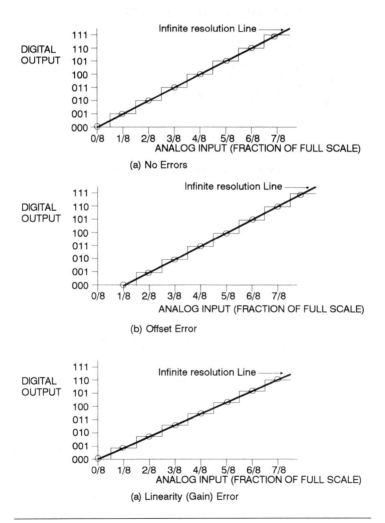

Figure 4-14 Three-bit ADC transfer curves illustrating errors.

An offset error would move the entire curve to the left or right, unchanged. This type of error can be corrected by adjusting the analog reference voltage. Figure 4-14b shows an offset error of 1 LSB.

A linearity or gain error would be equivalent to having the slope of the infinite resolution line vary, producing a larger error for larger input values. This would be more difficult to correct for, especially if it was temperature dependent. Figure 4-14c shows a linearity error of less than 1 (the gain drops at larger inputs).

Special-Purpose ADC Approaches   The ADC techniques discussed in this chapter have been standard, general-purpose approaches, in common use. Sometimes, a data acquisition system can be tailored to a special application for increased performance (hopefully without a significant cost penalty). One class of special applications particularly amenable to unique ADC systems is the realm of repetitive signals. These are identical waveforms that can be produced multiple times, without any significant change. Basically, these are static measurements under complete experimental control.

This type of repetitive system allows us to use an extremely high effective sampling rate based on a relatively slow ADC. Let us assume that the waveforms of interest have measurable energy up to 1 MHz. We need to sample at 2 MHz, which at reasonably high resolution (such as 12 bits) would require a very expensive ADC. We can get by with a high-resolution, slow (i.e., 1 kHz sample rate) ADC by adding a sample-and-hold (S/H) amplifier and a timing controller.

The idea here is to take one sample of the waveform for each repetition of the waveform. The S/H amp must be able to capture an analog voltage with a 500-nsec window (equivalent to a 2-MHz sample rate). The timing circuit must be able to step through the waveform in 500-nsec increments. For each repetition of the waveform, the next 500-nsec aperture is captured and digitized. The ADC's maximum conversion rate of 1 kHz determines the maximum waveform repetition rate. If the width of the waveform is 100 μsec, it would take 200 repetitions or 200 msec to sample it at effectively 2 MHz. See Chapter 14 for an example of this technique.

This survey of DACs and ADCs should help you decide which commercial hardware solutions are best suited to your data acquisition problems, or whether to build your own special purpose system.

CHAPTER

The Personal Computer

A computer is the heart of any contemporary data acquisition system. In the early 1980s minicomputers were the workhorse of most science and engineering labs. Hardware was expensive, most software had to be written in-house, and performance was barely adequate for all but the most expensive systems. Today, personal computers are commonplace throughout the scientific and engineering communities. The low cost and relatively high performance of personal computers makes them the ideal platform for most data acquisition tasks. In addition, a plethora of high-quality commercial software is available for all imaginable personal computer applications, including data acquisition and analysis.

Today's high-end engineering desktop computer is the workstation. This is typically a system with several megabytes (Mbytes) of volatile memory, a high-resolution video display, a relatively large amount of on-line storage (typically a hard-disk drive of over 100 Mbytes), a network connection, and a fast microprocessor (often a RISC CPU, or *reduced instruction set computer*). They are usually the platform of choice for a very high-performance data acquisition system at a relatively high price.

Even though workstations are clearly more powerful than standard personal computers, the distinction begins to blur when one looks at high-end personal computers. In fact, the major differences between a high-end personal computer and a low-end workstation are price and software availability.

Several popular classes of personal computers are useful as platforms for data acquisition systems. The ones we will examine in this book are based on the IBM PC/XT/AT bus, the IBM Micro Channel bus, and the NuBus used in Apple's Macintosh II systems. The IBM and compatible machines are based on Intel's 80×86 microprocessor (or CPU, *central processing unit*) family. The Macintosh machines are based on Motorola's 680×0 family of microprocessors. Since the most widely established class of personal computers use the IBM PC/XT/AT architecture, we will examine these machines in the greatest detail.

There are several members in Intel's 80×86 family. The original device was the 8086, a "true" 16-bit CPU. It had a 16-bit-wide data bus and a 20-bit address bus, providing for a 1-Mbyte address range. The original IBM PC and PC/XT used Intel's 8088 CPU, which was effectively an 8086 with only an 8-bit external data bus and a 20-bit address bus, for a 1-Mbyte address range, while keeping the same 16-bit registers internally for 8086 software compatibility. When the IBM PC was released in 1981, this hybrid approach of 16-bits internal and 8-bits external was common.

The IBM PC/AT used Intel's 80286 CPU, which employed a true 16-bit architecture, a 16-bit external data bus, and a 24-bit address bus, for a 16-Mbyte address range. It was software-compatible with the 8088 while providing faster processing speed and additional features. The expansion bus of the IBM PC/AT computer, a superset of the PC/XT expansion bus, eventually became an explicit standard: ISA (for *industry standard architecture*).

The next Intel processor was the 80386, which used a 32-bit architecture both internally and externally. It had a 32-bit external data bus and a 32-bit address bus, for a 4-gigabyte address range. IBM switched to its newer PS/2 line of PCs with the Micro Channel bus to use the 80386 and later CPUs. Many clone manufacturers stayed with the original AT (ISA) bus, with modifications for 32-bit-wide memory to accommodate 80386 machines. The ISA bus is still the common standard for 16-bit PC peripherals, even in 32-bit computers.

The latest Intel processor in this family (as of this writing) is the 80486. It is another 32-bit device with the same bus widths and features as the 80386 plus additional integrated functions, such as a floating-point processor. IBM is presently basing its high-end PS/2 systems on the 80486. Other manufacturers use it in ISA systems.

In a similar fashion, Apple's Macintosh computers have evolved along with Motorola's M68000 CPU family. The original member of this group was the 68000 with an external 16-bit data bus and a 24-bit address bus, providing a 16-Mbyte address range. Motorola also produced an external 8-bit CPU, the 68008, with a 20-bit or 22-bit address bus, for a 1-Mbyte or 4-Mbyte address range. This device is analogous to Intel's 8088 CPU. The 68020 CPU was a full 32-bit processor, with a 32-bit data bus and address bus, for a 4-gigabyte address range. The first Macintosh

II computers used the 68020. The next processor in Motorola's family was the 68030, another full 32-bit CPU with additional features to enhance performance, which is also used in the Macintosh II series.

Since the Macintosh II series has an open architecture (with publicly available specifications, unlike the earlier Macs) based on NuBus expansion slots, there is a growing number of add-in cards and support software for these personal computers. This includes the area of data acquisition. The Macintosh II computers are the only Apple systems we will cover in this book.

5.1 IBM PC/XT/AT and Compatible Computers

We will now look in-depth at the IBM PC/XT/AT class of PCs and their compatibles (sometimes called clones). First we will examine the IBM PC/XT computer, which is based on the Intel 8088 CPU. It has an external data bus eight bits wide and an address bus 20 bits wide, for an address range of 1 Mbyte.

5.1.1 Memory Segmentation

One idiosyncrasy of the 16-bit processors in this Intel CPU family is the way 20-bit physical addresses are generated from 16-bit registers. Intel uses an approach called segmentation. A special segment register specifies which 64-Kbyte section of the 1-Mbyte address space is being accessed by another 16-bit register. A segment register changes the memory address accessed 16 bits at a time, because its value is shifted left by four bits (or multiplied by 16) to cover the entire 20-bit address space. The segment register value is added to the addressing register's 16-bit value to produce the actual 20-bit memory address. Four segment registers and five addressing registers are available in an 8088, each 16 bits wide.

For example, when the stack is accessed, the 16-bit value in the Stack Segment (SS) register is shifted left by four bits (to produce a 20-bit value) and added to the 16-bit Stack Pointer (SP) register to get the full 20-bit physical address of the stack. The value added to the segment is referred to as the offset. The usual notation is *segment:offset*. So, if the code segment (CS) contained B021h and the instruction pointer (IP) contained 12C4h, the segmented notation is B021:12C4 and the physical location addressed would be B14D4h.

In contrast to this, the Motorola M68000 CPU family uses linear addressing, with internal address registers as wide as the physical address bus, so no segmentation is required.

Note that throughout this book, most addresses will be presented in *hexadecimal* (base 16) notation (with digits 0–9, A–F) using a trailing h. For example, 100h = 256 (decimal).

5.1.2 Motherboards

The heart of any PC/XT/AT computer is a single printed circuit board (PCB) referred to as the system board or the *motherboard*. It contains the CPU, some or all of the system's memory, timing, and control functions, as well as external interface capabilities (*input/output* or I/O). This external I/O is usually available through connectors on the motherboard, often referred to as expansion slots. Various cards are plugged into these slots, including display adapters (video controllers), disk drive controllers, parallel and serial interfaces, as well as boards for data acquisition.

5.2 The IBM PC/XT

A simplified block diagram of a PC/XT motherboard is shown in Figure 5-1. This motherboard contains the CPU, an optional coprocessor (Intel 8087) for floating-point math, eight hardware interrupts, four direct memory access (DMA) channels, three timer/counter channels, read/write memory (usually referred to as Random Access Memory, or RAM), Read Only Memory (ROM), all the required control logic, and interfaces to the external I/O slots. The 20-bit address bus, 8-bit data bus, and various control lines go to the I/O slots to support numerous peripherals.

Even though the 8088 can address 1 Mbyte of memory, only 640 Kbytes of RAM is usable on the PC/XT, in the address range 0 to 9FFFFh. The upper 360 Kbytes are reserved for system ROM and memory on expansion cards, which plug into the I/O expansion slots on the motherboard. A simplified PC/XT memory map is shown in Table 5-1.

5.2.1 I/O Addressing, Interrupts, DMA, and Timers

For communicating with peripheral, nonmemory (I/O) devices, the 8088 CPU supports both I/O mapped and memory-mapped I/O. I/O mapping separates I/O addressing from memory addressing, so I/O ports can be directly and easily accessed, even if they have the same addresses as memory locations. In memory mapping, I/O ports look like memory addresses and use up part of the memory addressing space. In the PC/XT design, I/O mapping is used. Although the 8088 will support 16 bits of I/O addressing, only 10 bits are used here (for a total of 1024 I/O addresses).

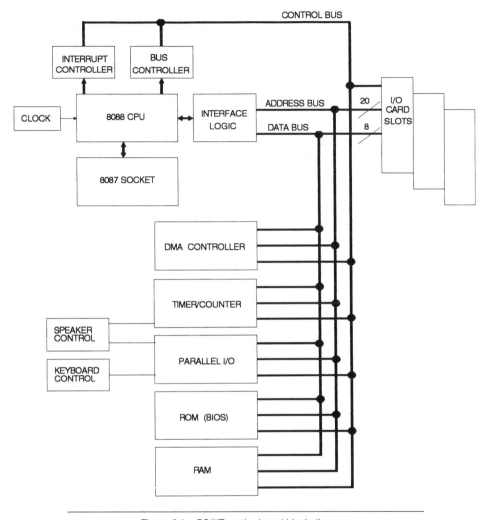

Figure 5-1 PC/XT motherboard block diagram.

This I/O space is divided into two regions of 512 locations each. The lower 512 addresses (0 to 1FFh) are used exclusively on the motherboard. The upper 512 addresses (200h to 3FFh) are decoded by interface cards connected to the I/O slots. An I/O address map for the PC/XT is shown in Table 5-2.

The PC/XT has nine interrupt lines or levels, with unique priorities. The highest priority interrupt is the NMI (nonmaskable interrupt), used

TABLE 5-1
PC/XT Memory Map

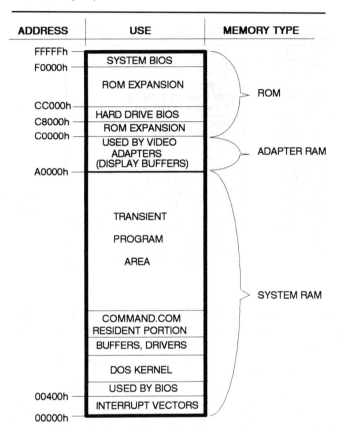

for trapping serious system problems such as memory (RAM) parity errors. The next two interrupts, IRQ0 and IRQ1, are also used only by the motherboard (IRQ1 interrupts the processor whenever the keyboard is hit). The other six interrupts, IRQ2–IRQ7 are available for use by cards in the external I/O slots. The lowest priority interrupt, IRQ7, is allocated to a parallel printer port.

Note that very often peripheral board manufacturers use interrupts in nonstandard ways for functions not previously defined. The same problem holds true for the use of I/O addresses and even memory addresses above 640 Kbytes (the limit of MS DOS). This is especially true for some PC/XT data acquisition cards. If two cards in the same PC try to use the same interrupt or address, they will malfunction. This is an incompatibil-

TABLE 5-2
PC/XT I/O Address Map

I/O ADDRESS	USE	
000 - 00Fh	DMA CONTROLLER	
020 - 021h	INTERRUPT CONTROLLER	
040 - 043h	TIMER	
060 - 063h	PPI (8255)	ON MOTHERBOARD
080 - 083h	DMA PAGE REGISTERS	
0A0	NMI MASK REGISTER	
200 - 20Fh	GAME ADAPTER	
210 - 217h	EXPANSION UNIT	
2F8 - 2FFh	ASYNCH ADAPTER (COM2:)	
300 - 31Fh	PROTOTYPE CARD	
320 - 32Fh	HARD DISK DRIVE ADAPTER	
378 - 37Fh	PRINTER ADAPTER	
380 -38Ch	SDLC COMM ADAPTER	ON ADAPTER CARDS
390 -393h	CLUSTER ADAPTER	
3A0 - 3A9h	BISYNC ADAPTER	
3B0 - 3BFh	MONO DISPLAY/PRINTER ADAPTER	
3D0 - 3DFh	CGA ADAPTER	
3F0 - 3F7h	DISKETTE DRIVE ADAPTER	
3F8 - 3FFh	ASYNCH ADAPTER (COM1:)	

ity, or an address clash. The solution is to change the interrupt/address selection on one or the other card, or remove one card entirely.

Another important PC/XT feature is the use of direct memory access (DMA). DMA hardware allows data to be transmitted very quickly between a peripheral device and system memory without the CPU's intervention. Programmed I/O transfers under CPU control are inherently much slower than DMA I/O transfers. DMA is especially useful for accessing diskette and hard disk drives. The CPU initializes the DMA controller with the required information and the DMA controller takes over the system bus, managing the data transfer.

There are four DMA channels in a PC/XT system. The highest-priority DMA channel (DMA channel 0) controls memory refresh, as discussed below. The other three DMA channels (1–3) are available for use by external I/O cards. Care must be taken in using DMA transfers, which can prevent normal CPU actions and result in a system crash.

The PC/XT contains three programmable timer/counters. The first timer/counter (channel 0) is implemented as a general-purpose time-of-day clock, producing a level 0 interrupt (IRQ0) approximately every 55 milliseconds. The second timer/counter (channel 1) times the DMA cycles for memory refresh, as described below. The third timer/counter (channel 2) controls the speaker's tone generation. If you need to use one of these timer/counters for other applications, try to use channel 2 only! This will not interfere with any critical system functions, whereas using other channels might.

5.2.2 PC/XT Memory, RAM and ROM

The PC/XT's main system memory consists of dynamic RAM. This read/write memory starts at address 0 and can extend up to 640K (9FFFFh). This is the memory used by the operating system, DOS, and is available for loading and running programs along with any transient data storage required by those programs.

Two types of RAM devices are static and dynamic. Both memories retain their contents only while power is applied to them. Dynamic RAM (DRAM), in addition, requires a periodic read access (on the order of every few milliseconds) to retain its memory. This process is called a refresh cycle. This is because each memory cell in a dynamic RAM acts like a capacitor whose charge slowly leaks off over time; it needs to be periodically recharged to the appropriate voltage.

Even though DRAM refresh uses up a finite amount of CPU time, it is commonly used in PCs because of its lower price-per-bit than static RAM and its higher density (more bits per package). When the original IBM PC appeared in 1981, its motherboard supported only 64 Kbytes of DRAM, using 16-Kbit ICs. Ten years later, 1-Mbit DRAMs are commodity items, 4-Mbit DRAMs are available, and 16-Mbit DRAMs are on the way.

The typical DRAM is 1 bit wide, so a 1-Mbit DRAM is configured as 1,048,576 (2^{20}) addresses by one bit. Most PC/XT/AT machines use nine DRAMs to produce a memory block one byte (eight bits) wide, with the additional bit used for *parity checking*. This is a hardware scheme to detect whether there was an error in reading memory. The DRAM refresh time on a PC/XT system can use approximately 7% of the available system time. This is accomplished using DMA channel 0 and timer channel 1.

The PC/XT's ROM contains the nonvolatile memory required to start up the system. This includes hardware initialization, power-on diagnostics (including a memory test), and a *bootstrap* program. The bootstrap allows the PC to load the operating system and start running it, usually from a diskette or hard disk drive. This allows for the flexibility to upgrade or even change the operating system a PC uses, without any hardware changes. Other important contents of the system ROMs include the programs needed for low-level control of various hardware I/O devices (such as disk drives, displays, keyboard). This is referred to as the basic input/output system, or BIOS (sometimes denoted ROM BIOS).

This *firmware* (software resident in a nonvolatile memory IC) is continuously used by the operating system for interfacing to all system I/O devices. If special system hardware is not supported by the BIOS, usually a special piece of software, called a driver, must be loaded into the operating system before the hardware can be used. An example of this would be an add-on optical disk drive.

The most common operating system used with PC/XT/AT computers is DOS (*disk operating system*), often specified as IBM DOS or MS DOS (for Microsoft, its developer). It is a single-user, single-task operating system with a limited memory usage of 640 Kbytes (see Chapter 7 for a more detailed discussion of DOS).

The system ROM is located in high memory addresses, above F4000h. Expansion cards plugged into the I/O sockets may also contain ROM, for integration into system code. This ROM may be present within the address range of C0000h–DFFFFh. If it contains valid information, the system will be able to execute the code (instructions) it contains. This is a common approach for hard-disk drive controllers or special video display adapters.

5.2.3 PC/XT Expansion Bus

The key to the PC/XT's flexibility is its expansion bus, with connectors for external I/O cards. Figure 5-2 shows the bus connections to an expansion slot. This bus gives an add-in card access to all the system address, data, and control lines except those dedicated to the motherboard, such as IRQ0, IRQ1, and DRQ0.

Here is a brief description of the I/O bus signal lines, designated pins A1–A31 and B1–B31 (as shown in Figure 5-2): Lines A0–A19 (pins A31–A12) are the address bits used for memory and I/O addressing, where A0 is the least significant bit (LSB) and A19 is the most significant bit (MSB). These are output lines relative to the motherboard. Similarly, signal lines D0–D7 (pins A9–A2) are the data bits, used for all data transfers (including DMA cycles), where D0 is the LSB and D7 is the MSB. These lines are bidirectional, both input and output.

Signals DRQ1–DRQ3 (pins B18, B6, B16) are the DMA request lines for channels 1–3. They are input lines, used by external devices to initiate a DMA cycle. Signals DACK0–DACK3 (pins B19, B17, B26, B15) are DMA acknowledge lines. They are outputs used to indicate DMA activity, acting as handshake signals for their respective DRQ lines.

Signals IRQ2–IRQ7 (pins B4, B25–B21) are interrupt request input lines, used by an external device to generate a CPU interrupt. IRQ2 is the

CHAPTER 5 The Personal Computer

	B	A	
B1	GND	-I/O CH CK	A1
	RESET DRV	D7	
	+5V	D6	
	IRQ2	D5	
	-5VDC	D4	
	DRQ2	D3	
	-12VDC	D2	
	Reserved	D1	
	+12VDC	D0	
B10	GND	I/O CH RDY	A10
	-MEMW	AEN	
	-MEMR	A19	
	-IOW	A18	
	-IOR	A17	
	-DACK3	A16	
	DRQ3	A15	
	-DACK1	A14	
	DRQ1	A13	
	-DACK0	A12	
B20	CLK	A11	A20
	IRQ7	A10	
	IRQ6	A9	
	IRQ5	A8	
	IRQ4	A7	
	IRQ3	A6	
	-DACK2	A5	
	T/C	A4	
	ALE	A3	
	+5VDC	A2	
	OSC	A1	
B31	GND	A0	A31

Figure 5-2 PC/XT I/O card slot connector.

highest priority and IRQ7 is the lowest. The system has to be properly initialized prior to an interrupt generation for it to be properly serviced.

Signal IOR (pin B14) is an output line indicating an I/O read cycle. This tells the external I/O device addressed to place its data on the bus. Similarly, IOW (pin B13) is an output signal indicating an I/O write cycle. This instructs an external I/O device to read data from the system bus. MEMR and MEMW (pins B12, B11) are the equivalent read and write output lines for reading from and writing to memory.

Signal I/O CH RDY (pin A10) is an important input line. It is used by slow memory or I/O devices to lengthen a read or write cycle. This is known as inserting *wait states*. It allows slower (and less expensive) peripherals to interface to the PC/XT, with only a penalty of more time required for a data transfer. If this signal is not used properly, it can be asserted for too long (more than a few microseconds) and effectively monopolize the system bus, preventing other activities. This could result in a system crash, where DRAM is not being properly refreshed or important interrupts are not being serviced. Figure 6-5, in Chapter 6, illustrates how to safely control I/O CH RDY.

Signal AEN (pin A11) is an output line used to prevent the CPU and other devices from accessing the system bus during DMA transfers. Signal ALE (pin B28) is an output line used to latch valid bus addresses by memory and peripheral devices. Signal I/O CH CK (pin A1) is an input line used to indicate a memory or I/O device parity error. Signal RESET DRV (pin B2) is an output line used to initialize (reset) devices on the bus at system power-on. Signal T/C (pin B27) is an output line that indicates when the maximum DMA transfer count is reached.

Signal OSC (pin B30) is an output line containing a 14.31818-MHz clock, with a 50% duty cycle. This clock may be divided down to provide other clock signals, such as dividing by 4 for the 3.58-MHz color video subcarrier frequency. On original PC and PC/XT systems, it was divided by 3 to provide the main system clock frequency of 4.77 MHz. Signal CLK (pin B20) is an output line containing the main system clock, with a 33% duty cycle. It is higher than 4.77 MHz in most contemporary PC/XT compatible systems. The most common clock frequencies used are 8 and 10 MHz. Obviously, the higher the system clock, the faster the CPU will operate. Overall system performance is not necessarily proportional to this clock frequency. In fact, some slower peripheral cards may not work properly with faster clocks, unless enough wait states are inserted.

The other lines on the I/O bus connector are power for the expansion cards. These lines are +5 V (pins B3, B29), −5 V (pin B5), +12 V (pin B9), −12 V (pin B7), and ground (pins B1, B10, B31). The positive voltage supplies typically have a higher current capability and are regulated to ±5%, as opposed to the negative supplies regulated to ±10% with lower current capacity. The original IBM PC's power supply could only produce approximately 65 watts of DC power, mostly for the +5-V (7 amps, maximum) and +12-V (2 amps, maximum) supplies. Most PC/XT compatible systems now use a power supply providing 120–150 watts of DC power.

For examples using some of these expansion bus signals, refer to Chapter 6.

5.3 The IBM PC/AT

Now we will examine IBM PC/AT computers and the ISA bus. The original IBM PC/AT and compatible systems were based on the Intel 80286 CPU. This was an expansion of the PC/XT architecture, including

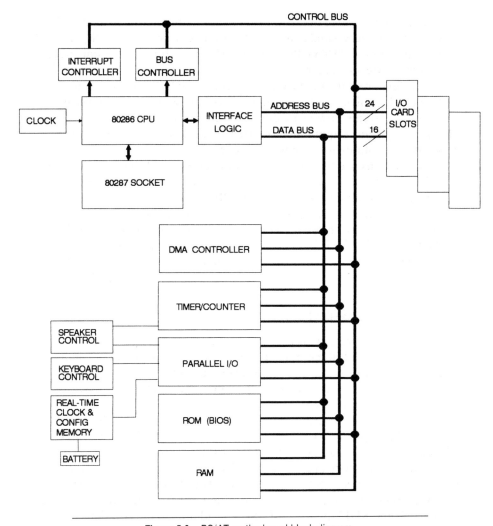

Figure 5-3 PC/AT motherboard block diagram.

the external I/O bus. The PC/AT block diagram is shown in Figure 5-3. The 80286 processor increases the number of address bits to 24, for a 16-Mbyte addressing space, and the number of data bits to 16. The motherboard now has 16 interrupt levels and seven DMA channels. It still has three timer/counters. New features include a real-time clock with battery-backup CMOS RAM. This small amount of memory stores clock and system configuration data.

5.3 The IBM PC/AT

The functioning of the IBM PC/AT (normally referred to as an AT or ISA system) is very similar to the PC/XT operation. Due to the higher performance of the 80286 CPU, overall system performance is enhanced. In addition, external data transfers can be 16 bits at a time, although 8-bit data transfers are still supported. The original IBM PC/AT had a 6-MHz system clock, which was later upgraded to 8 MHz. Most 80286-based AT compatible systems now use clocks ranging from 8 MHz up to 16 MHz. The faster systems require memory (RAM) with fast access time (or they must add wait states to memory access cycles).

AT systems use two connectors for each external I/O card slot. One is a 62-pin connector, compatible with the single PC/XT I/O connector. The differences are that now pin B4 is IRQ9 instead of IRQ2, pin B19 is REFRESH instead of DACK0, and previously unused pin B8 is now 0WS. Also, CLK (at pin B20) is faster and has a 50% duty cycle. Most cards designed for the PC/XT bus will work in an AT, as long as they can deal with the higher clock frequency and do not do any special remapping of memory.

5.3.1 PC/AT (ISA) Expansion Bus

As shown in Figure 5-4, AT I/O slots have a new, second connector consisting of 36 additional pins. These lines carry the additional address and data bits, IRQ signals, DMA signals, and special control lines that allow for 16-bit data transfers, zero wait state memory accesses, and multiple CPU operations.

Here is a brief description of these new I/O bus signals: Signal 0WS, added to the original 62-pin connector at pin B8, is an input line used to tell the CPU not to add any wait states to the present bus cycle. This is useful for fast memory and I/O cards. The remaining new signal lines are on the new 36-pin connector, designated C1–C18 and D1–D18. The additional address lines are LA17–LA23 (pins C8–C2). The additional data lines are SD08–SD15 (pins C11–C18). The additional interrupt lines available on the I/O bus (besides IRQ9) are IRQ10–IRQ12, IRQ14, and IRQ15 (pins D3–D7). The additional DMA channel-control signals now available are DRQ0 and DACK0, DRQ5–DRQ7, and DACK5–DACK7 (pins D8–D15).

Additional control lines also exist on the 36-pin connector. MEM CS16 (pin D1) is an input signal used to signify a 16-bit, one wait-state memory transfer. Similarly, pin I/O CS16 (pin D2) is an input signal indicating a 16-bit, one wait-state I/O data transfer. Signal SBHE (pin C1) is a bidirectional line used to indicate a data transfer on the upper eight bits (D8–D15) of the data bus. This line is used by devices that support 16-

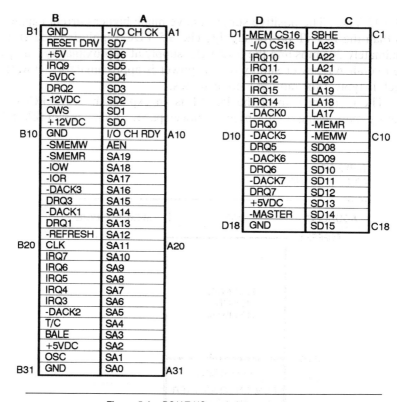

Figure 5-4 PC/AT I/O card slot connectors.

bit data transfers. Signal MASTER (pin D17) is an input line used by additional processors or DMA controllers to take control of the system bus. This line must be used carefully. If an external device holds the bus too long, system memory may be lost due to lack of DRAM refresh cycles.

Signal MEMR (pin C9) is similar to the original PC/XT bus signal MEMR (pin B12), now called SMEMR. The difference is, the original SMEMR is only active during a memory read cycle within the low 1 Mbyte of memory (original PC/XT address space). MEMR is active on all memory read cycles. Furthermore, SMEMR is an output line while MEMR can be either output or input. It can be driven by an external CPU. In a similar fashion, signal MEMW (pin C10) is a superset of the original MEMW (pin B11), now called SMEMW. The remaining lines on the 36-pin connector are extra power (+5 VDC) at pin D16 and ground at pin D18.

The PC/AT power supply provides +5 VDC, −5 VDC, +12 VDC,

and −12 VDC. The positive supplies have much higher current capabilities than the PC/XT power supply. The +5 VDC supply is rated at approximately 20 amps and the +12 VDC supply at approximately 7 amps. The overall AT power supply output power is approximately 200 watts, which is typical for most AT compatibles.

The memory map of the PC/AT is an expansion of the PC/XT's memory map, using a 16-Mbyte memory space, as shown in Table 5-3.

TABLE 5-3
PC/AT Memory Map

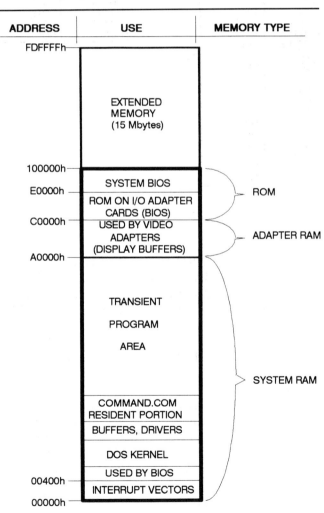

Note that the AT motherboard supports 64 Kbyte of ROM, as opposed to 40 Kbyte on the PC/XT motherboard. The PC/XT supports an Intel 8087 math coprocessor IC, for accelerated calculations involving floating-point math. The AT supports an Intel 80287 math coprocessor, for an 80286 CPU. If a system uses an 80386 CPU, it should support an 80387 coprocessor. Note that application software must explicitly utilize the math coprocessor for you to realize any benefit from it. Since these math ICs cost hundreds of dollars, they are typically installed only when required by a specific math-intensive application.

5.4 The BIOS

As mentioned above, the BIOS code located in ROM on a PC/XT/AT system handles the low-level software interface to the hardware. For example, to display a character on the video screen you send an appropriate command, along with the character, to the proper BIOS routine. Without the BIOS, you would have to know the intimate details of the video hardware, such as where in physical video memory to write the character for display. If the video display hardware is changed, software that directly addresses the hardware will no longer work. This is known as "ill-behaved" software. On the other hand, if BIOS calls were used, the BIOS will take care of hardware changes and the software can remain the same. This is "well-behaved" software. The penalty for using BIOS calls is a slower response than directly addressing hardware. Also, if a needed function does not exist in the BIOS, the hardware may need to be directly addressed. However, it is desirable to use BIOS functions whenever possible, as they will work universally with nearly all PC/XT/AT computers.

Some of the I/O facilities provided by BIOS routines support the keyboard, system clock/timer, communications ports, video display, floppy-disk drive, hard-disk drive, printer, system status, and ROM BASIC support (on true IBM systems only, not compatibles).

5.5 PC Peripherals

Nearly all PC systems use at least one floppy drive (a notable exception being diskless LAN workstations). Most XT and AT systems also have a hard-disk drive. It is strongly recommended that a PC-based data acquisition platform have at least a 20-Mbyte hard drive, for storage of raw and analyzed data as well as room for typically large application software.

Most PCs have at least one parallel printer port and a serial port, for asynchronous communications (see Chapter 8 for a discussion of parallel and serial interfaces).

Several standard video displays are available for PCs. The most basic is the text-only monochrome display using the monochrome display adapter (MDA). It offers one page of 25 lines of 80 characters with hardware support for high-intensity, underlining, and reverse video. It can support simple character-based graphics, where special characters are graphic symbols (such as lines) instead of alphanumerics. The MDA has a video buffer (memory) 4 Kbytes long. It produces sharp, easy-to-read text.

True bit-mapped color graphics are supported by the color graphics adapter (CGA). It provides four pages of 80-character by 25-line text, as well as several graphics modes. Its highest graphics resolution is 640 points horizontally by 200 points vertically in two colors. It also supports four colors with a resolution of 320 points horizontally by 200 points vertically. The CGA has a 16-Kbyte-long video buffer. Text on a CGA monitor is much "fuzzier" than on an MDA monitor. The original IBM PC only offered MDA and CGA display options.

The next available IBM video display is the enhanced graphics adapter (EGA). Its video buffer size varies from 64 to 256 Kbytes, and it can support multiple pages of text. It can display graphics with a resolution of 640 points horizontally by 350 points vertically, with up to 16 colors (with maximum buffer memory). It can also emulate a CGA or MDA display.

Some of the newer IBM PS/2 series of computers support multicolor graphics array (MCGA), which is an enhanced version of CGA. It uses 64 Kbytes of video buffer memory. It can store up to eight pages of monochrome text. For graphics, it supports all the CGA modes as well as adding support for 256 colors in a 320-points by 200-points mode. In addition, it has a high-resolution two-color graphics mode with 640 points horizontally by 480 points vertically.

A recent "standard" IBM video display for PCs is the virtual graphics array (VGA). VGA is used on many PS/2 systems as well as ISA systems. It has a 256-Kbyte video buffer. As does the EGA, it emulates MDA and CGA modes. It also emulates EGA and MCGA graphics modes. It can support a 640-point by 480-point high-resolution graphics display with 16 colors. VGA has become the most popular current display standard.

There is also one non-IBM video display standard, the Hercules graphics adapter (HGA), sometimes referred to as monochrome graphics. It was developed to fill the void between the original text-only MDA and color graphics CGA, as a graphics display using a monochrome monitor. It emulates MDA (and uses the same monitor) in text mode, along with MDA graphics characters. It can switch into a monochrome (two-color), *bit-mapped* graphics mode supporting a resolution of 720 points horizontally by 348 points vertically. Its video buffer contains 64 Kbytes of memory. Being a non-IBM standard, it is not supported by BIOS or DOS video functions. A special software driver must be installed to fully use it. However, many commercial software products support HGA, and it is a low-cost alternative to high-resolution color displays (EGA and VGA) when multicolor video is not required.

There are many other nonstandard video displays for PCs. Some of these enhance existing standard displays with additional, higher-resolution modes. Others are completely different and targeted at special applications. For example, many of the very high-resolution color displays (i.e., 1024 points by 1024 points with several colors) are useful for computer-aided design (CAD) systems running drafting applications. All of these nonstandard video systems require special software drivers and support from applications software. If you are configuring a system with a nonstandard display (or any other nonstandard device) make sure it is compatible with all the hardware peripherals and application software you intend to use. Be especially careful when a system uses local-area network (LAN) hardware and software.

The keyboard is the PC's standard user-input device, fully supported by BIOS and DOS functions. There are many other user input and control devices for PCs, the most popular being the mouse. The mouse is a device that connects to the PC via a serial port or a special I/O card. It is moved by the user's hand in a two-dimensional plane on an ordinary table top or a special pad. It has two or more buttons the user can push. In conjunction with supporting software, a mouse simplifies using graphics-based applications, such as CAD systems or desktop publishing systems. For example, a painting program allows the user to create and edit graphics images. A mouse can be used, among other things, to draw lines, select functions, and select objects on the screen to manipulate. Other, less popular peripherals for user input are digitizing pads and trackballs (a stationary version of a mouse, either built into a keyboard or free-standing).

An important, and sometimes overwhelming area of PC peripherals is that of mass storage. This includes floppy drives (diskettes), hard drives, and other esoteric storage devices. For floppy drives, two form factors are commonly used, 5-1/4 inch and 3-1/2 inch diskettes, each with two standard densities. The standard 5-1/4 inch drive supports double-sided double-density storage, which allows 360 Kbytes formatted capacity. This is common on XT class machines. Most AT machines use a double-sided high-density drive that is capable of 1.2 Mbytes of formatted storage. Similarly, both 3-1/2 inch drive formats are double sided. The double-density 3-1/2 inch drive has a formatted capacity of 720 Kbytes, and the quad-density drive has a 1.44-Mbyte capacity.

There are some wrinkles to note when using diskettes with different density drives. Most notably, if a diskette was formatted on a double-density 5-1/4 inch drive, it can be read by a high-density drive, but a high-density diskette cannot be read by a double-density drive. If a double-density diskette was written on by a high-density drive, sometimes it may not be read reliably by a double-density drive. Also, for both 5-1/4 and 3-1/2 inch drives, the diskettes used must be the appropriate type for that drive (so, do not use low-density diskettes in high-density drives and do not put high-density diskettes into low-density drives!).

The hard drive arena can be even more confusing. Hard disk drives can vary in capacity from 10 Mbytes to several hundred Mbytes. The most common sizes are in the range of 20 to 80 Mbytes. The majority of hard drives use MFM (modified frequency modulation) encoding. Some use RLL (run length limited) encoding to increase capacity and transfer speed by 50% over MFM. An important measure of performance is a drive's average access time, ranging from around 60 msec with lower-cost drives to under 20 msec on high-performance drives.

The type of drive-to-computer interface is another important hard disk parameter. The ST412 interface has been the *de facto* standard for most PCs. For very high performance drives, SCSI (small computer system interface) adapter cards are available for PCs (note that Macintosh computers come with a standard SCSI port). Many newer PCs have simple hard-disk interface electronics built into the motherboard. They interface directly to IDE (integrated drive electronics) hard drives without requiring an additional I/O card, as ST412 or SCSI drives do. A special I/O card can be used to interface to IDE drives if the motherboard lacks the capability. These drives are usually used with AT systems. IDE is quickly becoming the most common PC hard-drive interface standard. Some high-capacity (above 100 Mbyte) and high-speed (faster than 20 msec access time) hard drives for AT systems use ESDI (enhanced small-device interface) as an alternative to SCSI.

Another important class of mass storage devices are tape drives, typically used to back up data from hard drives. As users progress to larger hard drives, backing up their data onto diskettes becomes more cumbersome. For example, an AT system with a 40-Mbyte hard drive requires 33 high-density diskettes (1.2 Mbyte) for a total backup. Even using data compression techniques, about 20 diskettes would be required.

Instead, a tape drive using a single tape cartridge can store over 100 Mbytes. Tape drives are finally starting to show a trend toward standardization, making their use more attractive.

Optical drives are a young, rapidly evolving mass storage technology. The most common optical drive for a PC is the WORM (write once read many times) drive. As its name implies, you can only record on the media once, but read it back as often as needed. The media is replaceable and has very large capacity (from several hundred megabytes to a gigabyte). It is ideal for archiving data. Read-write optical drives are also appearing now, with standard hard drive capabilities. Their capacity is typically much higher than magnetic (hard disk) drives, but they tend to be much slower.

Similar to WORM drives, CD–ROM (compact disk–read only memory) drives are also available for PCs. These compact disks are prerecorded digital media (as are audio CDs) containing large amounts of reference information, such as an encyclopedia. As the name implies, they can only be read.

Other new mass-storage areas include a mixture of magnetic and optical storage. One example is the so-called "floptical" disk drive, which uses a floppy drive with optical tracks and a servo system for high-density storage. This opens the possibility for high-capacity diskette drives (around 20 Mbytes).

One final class of PC peripherals is that of printers and plotters. Most PC printers use either a parallel (Centronics) port or a serial port. Nearly all plotters use a serial port. A printer is used to produce text and graphics output. The majority of printers used are dot-matrix devices, forming characters and graphics images out of small, individual dots. Even laser printers use individual dots, albeit at very high densities (300 dots per inch or more).

Plotters are devices that produce drawings from a set of lines. They use one or more pens whose position on the paper is accurately controlled. Plotters are commonly used by CAD and graphic art software.

This completes our brief overview of XT and AT class PCs. In the next chapter we will look at the details of connecting external hardware to the PC/XT/AT I/O expansion bus.

CHAPTER

Interfacing Hardware to the PC Bus

We will now look at the details of connecting external hardware to an XT or AT bus. Initially we will examine 8-bit data transfers on a PC/XT bus. Later we will see the differences when connecting 16-bit devices to an AT (ISA) bus.

As we touched on in the previous chapter, three types of bus cycles are used for data transfers: memory, I/O port, and direct memory access (DMA) cycles. These can be either a read cycle, where data is transferred from an external device or memory into the CPU (or bus controller, when it is a DMA operation), or a write cycle where data is transferred from the CPU (or bus controller) to an external device or memory. Memory cycles are used to access system memory and memory on expansion cards (such as video buffers). Most data transfers to external devices use I/O port cycles or DMA cycles.

6.1 I/O Data Transfers

In XT systems, I/O port addresses in the range 200h–3FFh are available for use by I/O cards. Many of the I/O port addresses are reserved for particular functions. For example, the range 320h–32Fh is used by hard-disk drive adapter cards. One popular I/O address range for undefined functions is 300h–31Fh, assigned to IBM's prototype card.

Only a few control signals are needed, along with the address and data buses, to implement an I/O port read or write cycle. These are IOR (for a read cycle), IOW (for a write cycle), and AEN (to distinguish

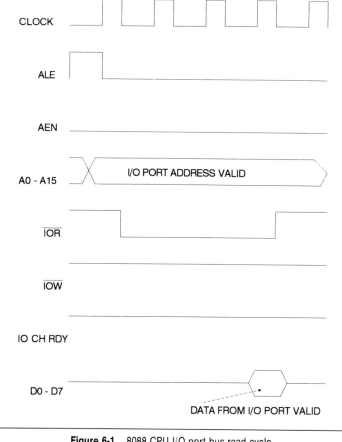

Figure 6-1 8088 CPU I/O port bus read cycle.

between an I/O port cycle and a DMA cycle). The timing for an I/O port read cycle is shown in Figure 6-1.

A standard PC/XT I/O port bus cycle requires five clock cycles, including one wait state injected by logic on the motherboard. Many systems with high clock frequencies inject additional wait states so that I/O cards designed for slower systems will still operate properly. The ALE signal occurs at the beginning of the I/O port cycle and indicates when the address bus contents are valid for the addressed port. IOR or IOW go active low to indicate an I/O port cycle. AEN stays inactive (low) to indicate this is not a DMA cycle. An active IOR signal tells the addressed I/O port to place its data (for the CPU to read) on the data bus (D0–D7). An active IOW signal tells the addressed I/O port to read the contents of the data bus (from the CPU). The control line I/O CH RDY is normally left active (high). If a slow I/O port needs additional wait states inserted into the cycle, it pulls this line low.

6.2 Memory Data Transfers

Memory bus cycles use timing very similar to I/O port bus cycles, as shown by the memory read cycle in Figure 6-2. The main control lines here are MEMR and MEMW. AEN is not needed for memory bus cycle decoding. One difference is that for memory bus cycles, the motherboard does not inject an additional wait state (hence, only four clock cycles are

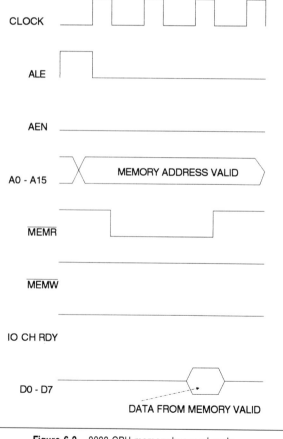

Figure 6-2 8088 CPU memory bus read cycle.

needed instead of five). Another difference is that all 20 address lines (A0–A19) are valid for a memory bus cycle and should be used for decoding the memory address. Only the first 16 lines (A0–A15) are valid for an I/O bus cycle; in practice, just the first 10 address lines (A0–A9) are decoded on a PC/XT bus.

6.3 A Simple 8-Bit I/O Port Design

A simple, fixed-address, 8-bit I/O port schematic is shown in Figure 6-3. The port I/O address is fixed at 300h by the decoding logic used on inputs A0–A9. IOW is used to write data to the output port latch (74LS373). IOR is used to read data at the input port buffer (74LS244). Note that the decode and control logic can be handled by a single PLD (programmable logic device) having at least 13 inputs and two outputs. A more versatile circuit would have a selectable I/O port address, determined by jumper or switch settings.

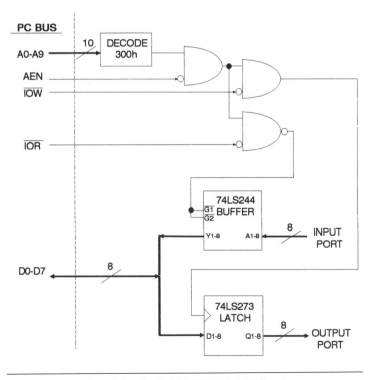

Figure 6-3 Simple 8-bit PC/XT digital I/O port.

Whenever the CPU writes to I/O address 300h, a data byte appears at the output port. When the CPU reads from that address, it retrieves the byte currently present at the input port. This is simple, programmed I/O that must be completely handled by the CPU. The CPU's program must determine when it is time for an I/O data transfer and must control the I/O read or write cycle as well as store or retrieve the data from memory. This limits the maximum data transfer rate and prevents the CPU from doing other tasks while it is waiting for another I/O cycle.

6.3.1 Using Hardware Interrupts

Usually, a better alternative to the polled I/O technique just described is to use hardware interrupts. The occurrence of a hardware interrupt causes the CPU to stop its current program execution and go to a special interrupt service routine, previously installed. This is designed to handle asynchronous external events without tying up the CPU's time in polling for the event. Nine hardware interrupts are used in a PC/XT system. The highest priority is the NMI (nonmaskable interrupt), which cannot be internally masked by the CPU (but can be masked by hardware on the motherboard). This line is usually used to report memory errors and is not available to cards connected to the I/O expansion slots. The other eight hardware interrupt lines, IRQ0–IRQ7, are connected to an Intel 8259 Interrupt Controller (which connects to the 8088's maskable interrupt input line). The highest priority lines, IRQ0 and IRQ1, are used on the motherboard only and are not connected to the I/O slots. IRQ0 is used by channel 0 of the timer/counter, and IRQ1 is used by the keyboard adapter circuit. Interrupts IRQ2–IRQ7 are available to I/O cards.

The 8088 CPU supports 256 unique interrupt types. These can be hardware or software interrupts. Each interrupt type has assigned to it a 4-byte block in low memory (0–3FFh) containing the starting address of that interrupt's service routine. This interrupt vector consists of the 16-bit code segment (CS) and instruction pointer (IP) of the service routine. Interrupt types 0–4 are used by the 8088 CPU. For example, interrupt type 0 is called by a divided-by-zero error. Interrupt types 5 and 6 are unused for 8088-based PCs. Interrupt type 7 is used by the BIOS for the Print Screen function.

Hardware interrupts IRQ0–7 are mapped to types 8–15. So, the vector for IRQ0 is at addresses 20h–23h, IRQ1 is at 24h–27h, and so on. A hardware interrupt is asserted when the appropriate IRQ line goes high and stays high until the interrupt is acknowledged. There is no direct interrupt acknowledge line from the I/O bus (it occurs between the CPU and the 8259 Interrupt Controller), so an I/O line under CPU control is used for this function and activated by the interrupt service routine.

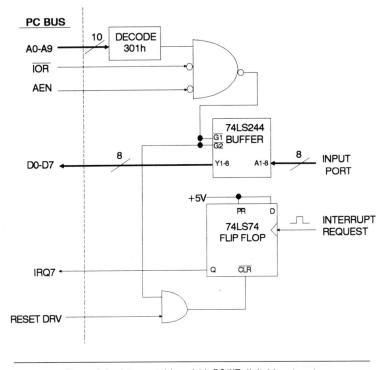

Figure 6-4 Interrupt-driven 8-bit PC/XT digital input port.

Figure 6-4 shows a simple 8-bit Input Port designed for interrupt-driven access, at I/O address 301h. As in Figure 6-3, the enable line of the input port buffer is decoded by a combination of address bits A0–A9, IOR, and AEN. In addition, the Input Port provides a Request for Interrupt line, used by the external hardware to signify when it is ready for the CPU to read data from it. A pulse or positive-going edge on this line sets the flip-flop, asserting the IRQ7 line (lowest priority interrupt). When the interrupt service routine for interrupt type 15 is called, it performs a read from I/O address 301h, to retrieve the data. This access will also reset the flip-flop, negating the IRQ7 line, preventing an additional (and unwanted) interrupt service cycle after the current one is completed.

Note that IRQ7 is typically used by a parallel printer port. To prevent unwanted hardware clashes, the flip-flop output in Figure 6-4 should be buffered by a tri-state driver, which can be disabled when the Input Port is not in use. A practical Input Port design would also have some selectability for the I/O port address and the IRQ line used.

Any interrupt type can be accessed via software by simply using the INT instruction. This includes interrupt types used by IRQ lines. This is a good way of testing hardware interrupt service routines.

6.3.2 Software Considerations for Hardware Interrupts

Implementing hardware interrupt support in software requires many steps. The interrupt service routine must be written and placed at a known memory location. The address of this service routine must be placed in the four bytes of low memory corresponding to the appropriate interrupt type (for IRQ7 it would be addresses 3Ch–3Fh). The 8259 Interrupt Controller must be initialized to enable the desired IRQ line. The 8088's maskable interrupt input must be unmasked (if it is not already). If you are using a standard peripheral device supported by BIOS functions, such as an asyncrhonous communications (serial) port, this initialization will be done for you by the BIOS. Similarly, commercial peripherals that come with their own software drivers should take care of these details for you. If you build your own data acquisition card with interrupt support, you will have to incorporate the initialization procedure into your custom software.

There are conditions where polled I/O is preferable to interrupt-driven I/O. It takes the CPU 61 clock periods to respond to a hardware interrupt and begin executing the interrupt service routine. In addition, it requires 32 more clock cycles to return from an interrupt. For an older PC/XT system with a 4.77-MHz clock, this corresponds to a processing overhead of 19.2 μsec added to the execution time of the interrupt service routine. If high-speed I/O transfers were required, such as every 20 μsec (for a 50,000 sample/second rate), a tight polling loop would be preferable. There would not be much time left over from servicing the I/O transfer for the CPU to do much else. In general, when the time between consecutive hardware interrupts starts approaching the overhead required to process an interrupt, a polled approach to software is in order.

6.4 DMA

When very high speed data transfers are required between a peripheral device and memory, direct memory access (DMA) hardware is often used. PC/XT systems support four DMA channels via an Intel 8237 DMA controller. The highest priority DMA is on channel 0, used only on the motherboard for DRAM refresh. The other three DMA channels are available for use by peripherals (channel 3 is the lowest priority). During a

DMA cycle, the 8237 takes over control of the bus from the 8088 and performs the data transfer between a peripheral and system memory. Even though the 8237 supports a burst mode, where many consecutive DMA cycles can occur, only a single-byte DMA cycle is used on PC/XT systems. This ensures that CPU cycles can still occur while DMA transfers take place, preserving system integrity (including memory refresh operations).

In PC/XT systems, DMA transfers require six clock periods. After each DMA cycle a CPU cycle of four clock periods occur. So, the maximum DMA transfer rate is a byte every 10 clock periods. On original 4.77-MHz PC/XT systems, this is every 2.1 μsec for a maximum DMA data rate of 476 Kbytes per second. This is still much faster than CPU-controlled data transfers.

As with servicing interrupt requests, software must perform initializations before DMA transfers can occur. The 8237 DMA controller must be programmed for the type of DMA cycle, including read or write, number of bytes to transfer, and the starting address. Once it has been properly initialized, the DMA cycle is started by a DMA request from the peripheral hardware.

6.5 Wait State Generation

As we previously discussed, sometimes a peripheral device is too slow for a normal PC/XT bus cycle. The length of a bus cycle can be extended by generating wait states, which are additional clock periods inserted into a memory or I/O bus cycle. This is done by pulling line IO CH RDY low (negated) for two or more clock cycles after the data transfer cycle has started.

Figure 6-5 shows a simple circuit for generating one additional wait state for an I/O cycle. When the I/O port is selected (for either a read or a write) it sets a flip-flop that pulls IO CH RDY low. Note that the inverter driving the IO CH RDY line is an open-collector device. This is because several peripherals on the PC/XT bus can drive this line simultaneously and will be OR-tied if they use open-collector outputs. This flip-flop output then goes to a two-stage shift register (using two additional flip-flops), which waits two clock cycles and then outputs a signal resetting the flip-flop and reasserting IO CH RDY, ensuring no additional wait states are injected into the cycle. For each additional wait state desired, an additional shift register stage should be added, for more clock cycle delays. The timing is very similar for generating memory cycle wait states, except only one clock cycle delay is required to generate the first wait state.

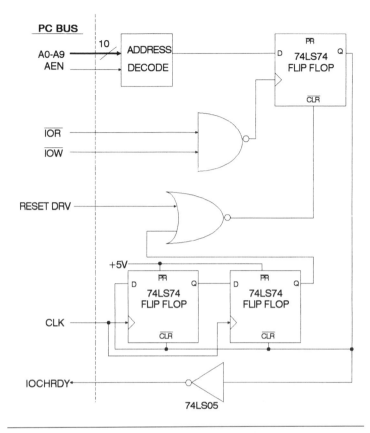

Figure 6-5 I/O wait state generation.

6.6 Analog Input Card Design

Building on what we have discussed in this chapter, Figure 6-6 shows an 8-bit data acquisition circuit with eight analog inputs. It is based on a National Semiconductor ADC0808 successive-approximation ADC with a maximum conversion rate of approximately 10,000 samples per second (100-μsec average conversion time). This device has an eight-channel analog multiplexer. It accepts input signals in the range of 0 to +5 V. If a wider analog input range is required, op amps can be used.

This circuit occupies I/O addresses 300h–307h. Writing dummy data to address 300h starts a conversion for the signal on ADC channel 1. A write to 301h converts channel 2, and so on. When conversion is complete, an IRQ is generated (the interrupt line used is jumper-selectable).

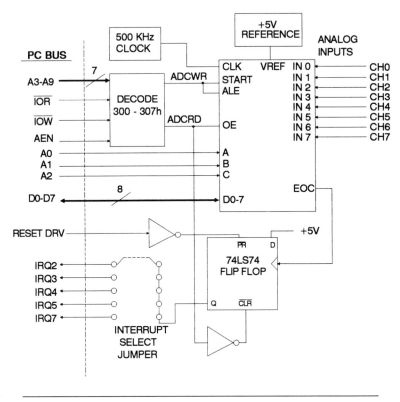

Figure 6-6 Eight-bit, 8-channel analog input card.

The interrupt service routine then reads the value from any I/O address in the 300h–307h range. The flip-flop that generates the IRQ is set by the ADC's end-of-conversion (EOC) signal and cleared when the interrupt service routine reads the ADC value.

6.7 16-Bit Data Transfers on ISA Computers

The PC/XT I/O circuits described above will also work in an AT (ISA) system. Most AT computers with high-frequency clocks (above 8 MHZ) insert additional wait states for I/O Port bus cycles so that cards designed for XT and slower AT systems will still work properly. Even 16-bit transfers to 8-bit peripherals are supported by hardware on the AT motherboard. However, to fully exploit the power of an AT system, an interface card system should support 16-bit data transfers wherever possible. This utilizes the additional data, address, and control lines of the AT I/O bus.

Basically, to perform 16-bit I/O port data transfers, we must decode the I/O port address, use IOR or IOW to determine the transfer direction, tell the system bus that we want a 16-bit transfer cycle, and input or output the 16-bit data word. An AT has the same I/O address map for devices connected to the system bus as the PC/XT (in the range 100h–

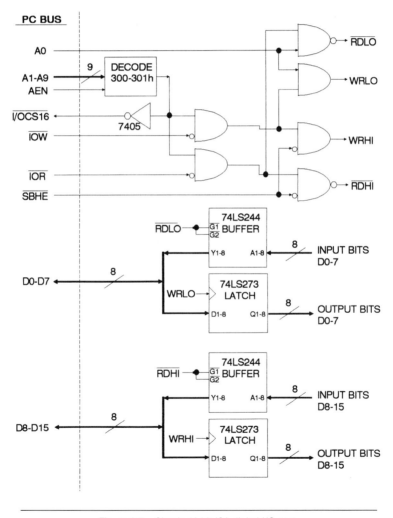

Figure 6-7 Simple 16-bit ISA digital I/O port.

3FFh). This makes I/O address coding the same. One new control line used on the ISA bus is I/O CS16 (pin D02), which indicates to the CPU (80286 or above) a 16-bit data transfer is requested by the peripheral device. Another new control line is SBHE (pin C1), which indicates when data on the upper byte of the data bus (D8–D15) is valid.

Figure 6-7 shows a simple 16-bit ISA I/O interface, designed for address 300h. The main difference between this circuit and the PC/XT I/O circuits shown previously is the transfer of 16 instead of eight bits at a time. Otherwise the I/O address decoding is the same, except for the LSB A0. In addition, the bus signal I/O CS16 is asserted, active low (by an open-collector driver), when the I/O port is addressed to request a 16-bit I/O transfer cycle. If this line was not asserted, as with an 8-bit PC/XT card, only the lower eight data bits (D0–D7) would be used for the I/O cycle. The signal SBHE is used when the upper eight data bits (D8–D15) are ready for bus transfer, and it enables the buffer for that data. A0 must be asserted to transfer the lower eight data bits.

It may be necessary, due to the higher clock frequency of most AT systems (especially 80386 and 80486-based computers), to add additional wait states to an I/O or memory bus cycle over and above the wait states automatically injected by logic on the motherboard. As with PC/XT systems, pulling the IO CH RDY line low can be used to add wait states to a bus cycle.

In the next chapter, we will examine software techniques for interfacing to personal computers. The topics covered will include how the PC's software system works and how to produce software to support peripheral hardware, especially for data acquisition applications.

CHAPTER

Interfacing Software to the PC

Using the correct techniques for interfacing software to a PC is as important as implementing the proper hardware interface. In this chapter we will start with an overview of PC/XT/AT software structure and then proceed to using this arrangement.

7.1 PC Software Layers

Four general layers of software are present on a PC, as shown in Figure 7-1. The lowest is the hardware level, where the software directly accesses the hardware. For example, if the addressed hardware was a display adapter, writing to a specific address in its video buffer (to display a character) would be directly accessing the hardware. At this level, the actual computer circuitry (I/O and memory addresses) determine the software instructions needed.

The next layer is the basic input output system, or BIOS. This is software, often referred to as firmware, residing in Read Only Memory (ROM) on the motherboard. The system ROM includes code to test the computer system and bootstrap (or boot) it, to begin normal DOS operation. The BIOS routines in ROM act as an interface between higher-level software and the actual hardware. They implement the details needed to operate various standard hardware peripherals (such as video displays or disk drives) and begin to provide some hardware independence. When a program uses a BIOS function, it does not need to know hardware-level details, such as the address of the status register on a disk drive controller

96   CHAPTER 7   Interfacing Software to the PC

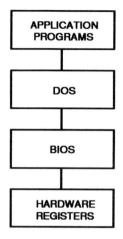

Figure 7-1   PC (MS-DOS) software layers.

card. It only needs to tell the BIOS the function it wants completed, such as to read data from a particular sector on a specified disk.

This hardware independence has important advantages. If different computers use different hardware components to carry out the same functions, this approach eliminates the need to rewrite a program for each machine, as long as the BIOS commands are the same. A hardware change in the same machine does not require a software change, as long as the BIOS supports the new hardware or is upgraded with it.

The only disadvantages with this approach are slower program execution and somewhat limited functionality. Since more instructions must be executed to produce a function from a BIOS call, compared to directly addressing hardware, a slower response is produced. For many functions, this is not important (such as the PC response when a user hits a key). When speed is required, such as in real-time control or data acquisition, direct hardware addressing may be necessary. If the BIOS functions do not support all the features of a particular hardware device, again direct hardware access may be required. Often, system software is loaded to supplement the BIOS and use the same software interface to call it, as described below.

The next layer of system software is the disk operating system, or DOS. This software is loaded into the PC's memory from a disk drive, by a bootstrap program in ROM. It operates at a higher level than the BIOS, even further removed from the hardware layer. Among other things, it implements the file and directory structure for disk drives. It advances the concept of hardware independence to device independence. For example, when a calling program requests data from a file under DOS, it does not need to know what type of physical drive contains the data. DOS keeps track of that information and retrieves the requested data by appropriate calls to BIOS functions. The program just uses a *logical* drive identification (such as A: or C:).

This device independence extends to the type of device, using the DOS feature of redirection, when it redirects data from one device to another. For example, the DOS TYPE command usually displays the contents of a text file on a video display (for example: TYPE MYDATA.TXT). DOS can redirect this data to a printer, with the command: TYPE MYDATA.TXT > PRN: (which sends this data to the system's default printer). A program calling DOS to perform these functions does not need to know about the differences between the two output devices (video display and printer) or even that very different BIOS calls are used to perform this function. DOS takes care of all these details.

The final, highest layer of PC software is the application program. This is the software that performs the useful functions we need a computer for in the first place, such as mathematical calculations, word processing, data acquisition, and graphical display. To perform these high-level activities, the application program calls various functions at the DOS, BIOS, and hardware levels. As before, for the highest degree of portability, maintainability, and hardware support, software interfacing should be at the highest level possible, preferably DOS, or BIOS if necessary. However, calling system functions through DOS is also the slowest route. As with BIOS calls, tradeoffs are sometimes necessary.

7.2 Software Interrupts

The mechanism for calling BIOS and DOS functions uses *software interrupts*. This provides a means of software independence for the called functions. A software interrupt works like a hardware-generated interrupt. It causes program execution to jump to a new location, specified by the interrupt number or level. There are 256 possible interrupt levels in 80×86-based PCs. Some are used by hardware interrupts, some by BIOS, and some by DOS. Table 7-1 lists that interrupt usage in a PC/XT system. To generate a software interrupt, the Assembler instruction INT, followed by the level (0–255), is executed. This specifies which interrupt vector to use. An interrupt vector is a four-byte address in low memory, 0–3FFh, which contains the location of the interrupt service routine. This

TABLE 7-1
Interrupt Usage in MS-DOS PCs

INTERRUPT #	CLASSIFICATION	FUNCTION
0 - 7	BIOS/DOS	CPU Interrupts
8 - F	BIOS	8259 Hardware Interrupts
10 - 1C	BIOS	BIOS Function Calls
1D - 1F	Data	Table Pointers (Video/Disk)
20 - 3F, 5C, 67	DOS	DOS Function Calls
80 - F0	BASIC	BASIC Functions

is the address the program jumps to when the interrupt is called, which contains the code to handle the interrupt request.

The beauty of this system is that the software calling the interrupt routine, such as a BIOS function call, does not have to know exactly where in memory the interrupt service routine is. This is the software independence alluded to above. If the BIOS code is upgraded at some future point, the absolute location of the interrupt service routine may change, but the software calling it does not have to change, since the interrupt vectors will also be upgraded.

7.2.1 BIOS Interrupts

Using a previous example, the BIOS routine interfacing with the video display works through INT 10h. To display an alphanumeric character on the current video screen, the character byte is loaded into CPU register AL (the low byte of the accumulator) and 14 is loaded into AH (the accumulator's high byte), which specifies the video command (display a character). Then an INT 10h instruction is executed. Written in Assembler, the code to display the character "9" would be

```
MOV AL, 39H
MOV AH, 14
INT  10H
```

Note that 39H is the ASCII code for the character "9."

As shown in this example, BIOS functions use some of the CPU's registers for sending data to and receiving data from the function called.

Sometimes, the Carry flag is returned to specify a particular condition. When one interrupt is used for several different functions (as Int 10h, 13h, 14h, 15h, 16h, 17h, and 1Ah), register AH is loaded with the function number. Table 7-2 is a summary of most of the BIOS functions available on PC/XT/AT systems.

7.2.2 DOS Interrupts

DOS functions are called by software interrupts similar to BIOS functions. Most DOS functions are called via INT 21h. DOS reserves the use of INT 20h–3Fh, although only INT 20h–27h are used for most common functions. Again, the function number is selected by the value placed in register AH. Some DOS INT 21h functions also have a subfunction, selected by the value in register AL.

As an example of using a DOS function, we will once again write a character to the video display, using INT 21h, Function 2. Here, register

TABLE 7-2
Standard MS-DOS PC Bios Functions

INTERRUPT #	PURPOSE
10 h	Video Display Functions (func 0-13h)
11 h	Equipment Check
12 h	Memory Size Check
13 h	Floppy Disk Functions (func 0-18h)
14 h	Communications Functions (func 0-5h)
15 h	Cassette and Misc System Functions (func 0-C4h)
16 h	Keyboard Functions (func 0-12h)
17 h	Printer Functions (func 0-2h)
18 h	Execute IBM BASIC from ROM
19 h	Re-Boot System
1A h	System Timer/Clock Functions (func 0-7h)
1B h	Keyboard CTRL-BREAK Interrupt Handler
1C h	System Timer Tick (18/sec) Interrupt Handler

AH contains the function number (2) and register DL contains the character to be displayed. If we use Microsoft C instead of Assembler in this case, we can write a general-purpose subroutine for video display called disp_ch( ):

```
#include <dos.h>              /* standard definition files */
#include <stdio.h>
#define FUNCT 2               /* function number 2 */

disp_ch(ch)                   /* subroutine name */

char ch;                      /* character argument */
{                             /* start of subroutine */
    union REGS regs;          /* sets up register use */
    regs.h.ah = FUNCT:        /* AH = 2 */
    regs.h.dl = ch;           /* DL = character to display */
    intdos(&regs,&regs);      /* call INT 21h */
}
```

A calling program, to display the character "9" would be

```
main()
{                             /* start of program */
    char c;
    c = 0x39;                 /* ASCII value for  9  */
    disp_ch(c);               /* call subroutine */
}
```

Even though more coding (along with more software overhead) is required to implement this DOS function in C, compared to Assembler, this approach is usually preferable. C is a high-level language with good functionality and ease of use. It is much easier to maintain a program in C than in Assembler, and the penalty of larger, slower programs is not as severe as with other high-level programming languages. We will discuss the various tradeoffs between different programming languages later in Chapter 13.

7.3 Polled versus Interrupt-Driven Software

In Chapter 6 we looked at the tradeoffs between accessing a peripheral device via polled software versus interrupt-driven software. If a peripheral device needs to be serviced relatively infrequently (for example, using only 10% of the available CPU time) and *asyncrhonously* (so the program cannot predict when the next service will be required), interrupt-driven software is in order. On the other hand, if interrupt servicing takes up too much CPU time (sometimes referred to as CPU bandwidth) for very frequent servicing, polled software would be preferable. In this case, there would be little CPU bandwidth left over for other processing anyway. One other general case is when the peripheral servicing is *synchronous*, as when the value of an ADC is read at preset time intervals and requires a small amount of CPU bandwidth. Again, interrupt-driven software is the best solution. If the peripheral (ADC) does not provide a hardware interrupt, the PC's timer could.

The following program listing, written in Microsoft Macro Assembler, shows the basic concepts for installing and using interrupt-driven software. It can be used with the data acquisition circuit from Chapter 6 (Figure 6-6), set to generate an IRQ7 hardware interrupt whenever a new ADC reading is ready. It is assumed that the 8259 interrupt controller already enables IRQ7 interrupts and that the system interrupt flag is set to enable the maskable interrupt input from the 8259. Otherwise, these functions must be taken care of in LOADVEC, the routine that prepares the system for the interrupt and loads the interrupt service routine INT7SVC, as follows:

```
; *** MACRO ASSEMBLER PROGRAM TO READ ADC VALUE VIA IRQ7 ***
;
;* DATA INITIALIZATION *
DSEG1      SEGMENT AT 0    ;interrupt vector table starts at addr 0
           ORG  3CH         ;start of vector for IRQ7
IRQ7       LABEL    WORD   ;Now we can access the vector for IRQ7
DSEG1      ENDS             ;via the label IRQ7
;
DSEG2      SEGMENT          ;Data storage segment
      PUBLIC    DVALUES, DINDEX    ;Allows other programs access
                                   ;to these variables.
DVALUES    DB 256 DUP (?)  ;ADC data storage table (uninitialized)
DINDEX     DW 0             ;Index into table (initialized to zero)
DSEG2      ENDS
;
CSEG       SEGMENT          ;Code segment, for programs
           ASSUME    CS:CSEG, DS:DSEG2
;
;* ROUTINE TO INITIALIZE IRQ7 & LOAD SERVICE ROUTINE INTO MEMORY
LOADVEC:   MOV AX,0         ;Point to memory segment 0 for interrupt
           MOV ES,AX        ;vector table.
           MOV ES:IRQ7,OFFSET INT7SVC    ;Set address of IRQ7
           MOV ES:IRQ7+2,SEG INT7SVC     ;service routine.
           MOV DX,200       ;DX contains amount of memory to save
                            ;for keeping service routine INT7SVC
                            ;loaded in memory.
           MOV AL,0
           MOV AH,31H       ;Get ready for DOS function 31h
           INT 21H          ;Return to DOS, leaving INT7SVC memory
                            ;resident.
;
;* INTERRUPT SERVICE ROUTINE
ADC        EQU  300H        ;Address of ADC port (to read data)
```

```
INT7SVC:    PUSH AX              ;Save all working registers
            PUSH DS
            PUSH BX
            PUSH CX
            PUSH SI
            MOV AX,DSEG2         ;Point to data storage segment
            MOV DS,AX
            IN AL,ADC            ;Read data from ADC
            MOV SI,DVALUES
            MOV [SI+DINDEX],AL   ;Store data in table
            INC DINDEX           ;Point to next location in table
            CMP DINDEX,257       ;Past end of table?
            JNZ CONTIN           ;No
            DEC DINDEX           ;Yes, stay at end of data table
CONTIN:     MOV AL,20H           ;Send EOI command to 8259
            OUT 20H,AL
            POP SI               ;Restore working registers before
            POP CX               ;returning.
            POP BX
            POP DS
            POP AX
            IRET                 ;Return from interrupt
;
CSEG        ENDS
            END    LOADVEC      ;Start execution at routine LOADVEC
;* END OF PROGRAM
```

Since IRQ7 is interrupt type 0Fh, its vector is located at memory address 0Fh * 4 = 3Ch in segment zero (physical address 0000:003Ch). When the program is run by DOS, it starts execution at routine LOADVEC. This short program loads the address of the interrupt service routine, INT7SVC, into the vector location for IRQ7 (3Ch–3Fh). Then it allocates enough space for INT7SVC and its data and returns to DOS, leaving INT7SVC resident in memory. This type of software is called terminate-and-stay-resident, or TSR. It is useful here, allowing the servicing of the IRQ7 interrupt independent of other software. The DOS call, INT 21h Function 31h, is used to load TSR programs. The value in DX is the amount of memory to preserve for the resident program. AL contains the value returned by the function, which is useful for error codes. AH contains the function number.

Once INT7SVC is loaded into memory, whenever it is called it reads the current value from the ADC and stores it in a data table, starting at location DVALUES and indexed by DINDEX. Both DVALUES and DINDEX are declared *Public* labels, so that other software can access them and retrieve the data. A typical program making use of INT7SVC would check the value in DINDEX, address the ADC, start a data conversion, and then go about other business. When it was ready to retrieve the data, it would check that DINDEX has incremented and then read the data out of the table DVALUES. When it was done, it would decrement DINDEX.

Note that the above program is merely an illustrative example of the use of interrupt-driven software for data acquisition. It is still fairly rough and incomplete for very practical use, without refinements. INT7SVC does show some important aspects of interrupt service routines. They should be as fast as possible, to avoid interfering with other system interrupts. That is why they are usually written in Assembler (although short C programs are sometimes used). The working system registers (AX, BX, CX, DS, SI) should be saved by PUSHing onto the stack at the routine's start and restored, by POPping, at its end. Otherwise, any use of these registers by the interrupt service routine will corrupt the interrupted program, on return. For hardware interrupt service, the routine must send an EOI command to the 8259 interrupt controller. Otherwise, new hardware interrupts will not be enabled. The service routine should end with an IRET statement for a proper return from the interrupt.

An interrupt routine to service a software interrupt is somewhat simpler, since the 8259 does not have to be serviced and hardware interrupts do not need to be unmasked. In addition, there is little danger of monopolizing the CPU's bandwidth (unless hardware interrupts are masked off). Software interrupts are a convenient way to install and call software functions in memory.

To illustrate polled software used to retrieve an ADC value, the following is a function written in Microsoft C:

```
#include    <conio.h>       /* needed for library function inp() */
#define     ADC_STATUS    0×301    /* Address of ADC status port */
#define     ADC_DATA      0×300    /* Address of ADC data port */

char read_adc()     /* Name of function is adc_read */
    {
    while (inp(ADC_STATUS) != 1);  /*wait till ADC is done */
    return(inp(ADC_DATA));         /* send ADC value back to
                                      calling program */
    }               /* Done */
```

Note that this is a very short and simple subroutine. The main program would call it whenever it has started an ADC conversion and wants to retrieve the results. It is assumed that I/O port 301H contains a value of 1 only when the conversion is complete. This is the status required by a polling routine like read_adc( ).

In this simple example, there is no provision for a situation wherein something goes wrong and the ADC status port never returns a 1, as when there is a hardware failure or a software bug calling read_adc( ) at the wrong time. A more practical program would have a time-out provision in the while (...) statement. Otherwise, the PC would remain stuck in that loop.

7.4 Device Drivers

Previously, we have seen how useful interrupts are, both for calling existing DOS and BIOS functions and for interfacing to additional software functions, especially to support hardware such as data acquisition devices. Another special type of software is the device driver. A device driver is a distinctive program that is loaded into DOS when the system boots up and then acts as if it is part of DOS. As such, it must adhere to very strict guidelines. Device drivers are typically used to support special hardware functions. For example, a hardware mouse will usually have a device driver that allows it to work with common application software packages.

In DOS, device drivers are loaded into the system by including commands in a text file called CONFIG.SYS in the root directory of the boot disk. This file contains entries used to customize DOS, such as number of buffers and number of files that can be open simultaneously. It also contains entries in the form:

```
DEVICE = filename
```

where filename is the name of a device driver, typically with a SYS extension. So, to load a mouse driver (file MOUSE.SYS), CONFIG.SYS should contain the line:

```
DEVICE = MOUSE.SYS
```

When DOS boots up, it looks for CONFIG.SYS and, if it is found, it executes the commands it contains and loads the device drivers listed in the file. It should be noted that DOS device drivers must be written in Assembler for the proper control of program and data layout. They are normally only written by experienced DOS programmers.

7.5 TSR Programs

When DOS software support is required for special hardware, often writing a terminate-and-stay-resident (TSR) program is an appropriate choice, especially if it is not for commercial product support. It is much easier than producing a device driver, and it can be written in a high-level language, like C.

As we previously touched on, a TSR program is interrupt-driven software. It is loaded into a PC's memory and can interface with other programs or with DOS itself. It continues to function until the system is turned off and RAM contents are lost. All TSR programs are activated by interrupts, either hardware or software. Some use software interrupt levels not reserved by DOS or BIOS, to allow an application program to access the TSR functions.

It is common for TSR programs to attach themselves to interrupts already in use. For example, many utility TSR functions are activated when a special combination of keys is pressed (a *hot key*). To do this, the TSR program attaches itself to the keyboard interrupt 09h. This interrupt occurs whenever any key combination is pressed. If the TSR program's hot key is pressed, it can take over and perform its function. If not, it passes control on to the original interrupt service routine. This is also an example of how interrupt routines can be chained, with more than one service routine using the same interrupt level. In a similar fashion, some TSR programs that must perform a task periodically use the system timer interrupt.

7.6 DOS

As the primary hardware focus of this book is on IBM PC/XT/AT systems and compatibles, the software focus is on Microsoft/IBM DOS as the operating system for these PCs. DOS is by far the most popular software environment used by 80×86-based PCs, but not the only one. It is a *single-user*, *single-task* operating system, meaning it can only do one thing (execute one program) at a time. For the majority of PC applications, especially data acquisition and control, this is adequate. For cases where mainframe style functioning is needed (such as multiuser support) a more sophisticated operating system could be used. Similarly, special operating systems are used for operating a local area network (LAN) connecting multiple PCs together.

DOS has grown considerably since its initial release in 1981. Version 1.0 for the original IBM PC only supported single-sided 5-1/4 inch floppy disks. Version 1.1 supported double-sided 5-1/4 inch floppy disks. Version 2.0 was released with the IBM PC/XT and added support for a hard disk drive. Version 2.1 added support for IBM's portable PC and its ill-fated PC*jr*. Version 3.0 was released for the IBM PC/AT and supported high-density (1.2 MByte) 5-1/4 inch floppy disks. Version 3.1 added networking support. Version 3.2 added support for 3-1/2 inch floppy disks. Version 3.3 included support for the new IBM PS/2 systems. Version 4.01 added expanded memory support and an optional, menu-based interface shell, enhancing its standard command-line interface. In addition, it allowed large disk drives (over 32 Mbytes) to be used as a single logical device. DOS versions below 4.0 required a hard disk greater than 32 Mbytes to be partitioned into multiple logical drives.

The primary advantage for using DOS is that it is supported by a vast array of commercial software products. In addition, it is relatively inexpensive. Its primary disadvantage, besides being a single-task environment, is its memory limitation. A DOS application can only directly access up to 640 Kbytes of system RAM, regardless of the hardware capabilities of the PC. This stems from the original PC's 8088 CPU with 1 Mbyte of physical addressing space available and 384 Kbytes reserved for memory on peripheral devices (such as video display and disk controller cards). As an additional limitation, DOS allocates some memory for its own uses, typically leaving well under 600 Kbytes available for use by an application program. In general, each successive version of DOS monopolizes more memory for itself.

When an 80286 or higher CPU (80386, 80486) runs DOS with its 1 Mbyte addressing limit, it is working in the processor's real mode, which is fully compatible with the 8088. To access physical memory above 1 Mbyte, the CPU must use its protected mode, which is not supported by DOS.

For many applications, the 640-Kbyte limit of DOS is not a problem. For data acquisition applications, however, this can be a severe limitation, especially when a huge amount of data is being acquired and analyzed. For example, let us assume a system was acquiring 16-bit data at a rate of 50,000 samples/second, running a program under DOS, and it had available 512 Kbytes of memory as data storage (the rest of the DOS range was needed for the program code). It would take just 5.12 seconds of data to fill up this memory buffer. Obviously, if more data acquisition was required for each test the data would have to be stored in a disk file as quickly as possible, before the memory buffer filled completely. If this data was being analyzed, the application program would have to keep reading in new data from the disk file if more than 5.12 seconds was stored. There are several ways to get around the memory limitations of DOS.

7.7 Non-DOS Operating Systems and Software Environments

To make full use of AT systems that can physically address more than 640 Kbytes of system memory (using 80286, 80386, or 80486 CPUs), special software or another operating system is needed to operate in the processor's protected mode. One such operating system from IBM and Microsoft, used instead of DOS, is OS/2. It allows a system to run large application programs using more than 640 Kbytes of RAM for code and data. It enables the use of an AT systems's *extended memory*, which starts at address 100000h or 10000:0000h (which is 1 Mbyte). Of course, an application must be compatible with OS/2 to make use of all the available extended memory.

Microsoft's Windows is an additional option for ATs. It is another software environment that supports large applications and makes full use of a system's physical memory. It is a multitasking environment, allowing more than one program to reside in memory and operate at any given time. Each program has its own window on the display screen. In addition, data from one program or window can be easily sent to another program or window, facilitating complex tasks using multiple applications (such as incorporating the results of a calculation into a document). It included a graphics-based user interface, analogous to Apple's Macintosh operating system. To take advantage of all these features, an application must be specifically written to be compatible with Windows.

Another software product, DESQview, from Quarterdeck, is a multitasking operating environment, similar to Windows. One other operating system we should note here is Unix. This is a multitasking, multiuser operating system developed for minicomputers by AT&T Bell Laboratories. It has been *ported* to (adapted for use on) many different computing platforms. It is especially popular on workstations and high-end personal computers. Standard Unix has a command-driven user interface, as DOS does. Its commands are terse and often cryptic. It does provide a large amount of power and flexibility. Some versions of Unix for use on PCs include a user-friendly interface, either menu-based or graphics-based. One common PC version of Unix is Xenix.

A disadvantage of the operating systems and software environments that work in an AT's protected mode is that they require a relatively large amount of system memory to make use of their full functionality (which can be 2 Mbytes or more). These products should only be used when multitasking or protected mode operation is a necessity. For typical data acquisition applications, their sole benefit is access to large amounts of memory. There are other options, while staying in a DOS environment, for working with large amounts of memory.

7.8 Overcoming DOS Memory Limitations

7.8.1 Overlays

When writing your own programs, a simple technique to reduce the amount of memory required for execution is to use overlays. An *overlay* is a section of program code that is loaded into memory only when

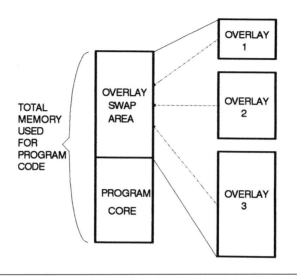

Figure 7-2 Example of program overlays.

needed; otherwise, it resides in a disk file. As illustrated graphically in Figure 7-2, an executable program residing in memory can consist of several code sections. These code sections, containing the program's instructions, can be subdivided into a program core, which is always resident, and one or more overlay sections. An overlay section contains code that can be swapped out and replaced by other code as the program executes. This swapping is controlled by the program core, which would contain all the functions and variables required by the various overlays. It is important for the individual overlay code sections to operate independently of each other, though not of the program core.

In the example of Figure 7-2, one overlay swap area is shared by three overlay sections. The overlay swap area must be as large as the biggest overlay that uses it. In this case, if the largest overlay is number 3, the memory saved by this technique (presumably for data storage), is the sum of the memory required for overlays 1 and 2. Of course, there are limitations on the amount of memory savings produced by using overlays, and a program's structure must be very carefully worked out to use them. One major drawback to using overlays is slow program execution. Every time an overlay is swapped into memory (from a disk drive) the program must wait. The more overlays a program uses, the more swapping will occur during execution and the slower the overall program will run.

7.8.2 Expanded Memory

One popular and well-supported technique for stretching the 640-Kbytes memory limit of DOS is called expanded memory, which should not be confused with an AT's extended memory (beginning at an address of 1 Mbyte). Expanded memory is a standard supported by Lotus, Intel, and Microsoft, referred to as the LIM standard, which provides access up to 8

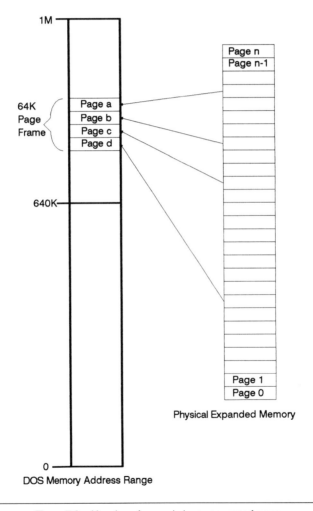

Figure 7-3 Mapping of expanded memory page frames.

Mbytes of extra memory, even on a PC/XT system. Expanded memory works within the 1-Mbyte DOS addressing range. It is a memory page-swapping technique. As shown in Figure 7-3, an unused block of memory up to 64 Kbytes long, between 640 K and 1M, is set aside as a page frame. This area can contain up to four pages of memory, each 16 Kbytes long. Special hardware (either a separate peripheral card or part of the system's motherboard) contains the physical memory storage: up to 8 Mbytes of pages, 16 Kbytes long. At any time, up to four pages of physical memory can be mapped into the 64-Kbytes page frame, where they are addressable by DOS and the rest of the system.

To make use of expanded-memory hardware a device driver must be installed into the system's CONFIG.SYS file. This driver is usually called EMM.SYS (for Expanded Memory Manager) and it operates through INT 67h. This driver controls the memory page mapping and allocation functions. Many applications support expanded memory when it is present in a system.

It should be noted that expanded memory is normally used just for data storage, since you cannot execute code from it or even from the page-frame space (above 640 Kbytes). LIM version 4.0 did add support for enhanced expanded memory, which can swap an entire program into and out of expanded memory and supports a multitasking environment.

Since the memory page mapping of expanded memory is controlled by dedicated hardware it is relatively fast, though not as fast as directly addressing memory in an AT system's protected mode (as long as there is no context switching between protected mode and real mode, which is fairly slow). Expanded memory is extremely useful for data acquisition applications that require large amounts (megabytes) of data storage in RAM, at data transfer rates that would outrun disk drive speeds.

7.9 Software Support for a Mouse

Before leaving the topic of PC software interfaces, one additional non-standard device should be mentioned: the mouse. As previously noted, a computer mouse is an input device that produces pulses when it is moved around a desktop. These pulses correspond to motion in the x and y directions that are translated to cursor motion on a video screen. In addition, a mouse has two or three buttons activated by the user to select some item or initiate a process. They are mostly used in graphics applications, such as CAD programs.

The most common software support for mouse functions is through a device driver (Microsoft's driver is MOUSE.SYS). The mouse functions are commonly called via INT 33h. The driver continuously updates the mouse cursor's position on the screen as the mouse moves, without requiring action from the application program using it. Employing INT 33h, the application program has complete control over mouse parameters.

This completes our survey of topics related to PC software interfacing. In the next chapter, we will explore common personal computer hardware interface standards, including GPIB and RS-232C.

CHAPTER

Standard Hardware Interfaces

Previously we saw how a Personal Computer's I/O operates from its expansion bus. However, not all external I/O goes directly through the expansion bus. Very often a standard hardware interface is used, by either another computer or an external peripheral device. We will explore several of these parallel and serial computer interfaces.

8.1 Parallel versus Serial Digital Interfaces

In general, digital computer interfaces to the outside world fall into two categories: parallel and serial. The differentiation between the two is important. For a digital interface n bits wide a parallel device uses n wires to simultaneously transfer the data in one cycle, whereas a serial device uses one wire to transfer the same data in n cycles. All things being equal (which they rarely are), the parallel interface transfers data n times faster than the serial interface.

Figure 8-1 shows an 8-bit-wide interface between a PC and an external device. For simplicity, let us assume the data is unidirectional. The parallel interface in Figure 8-1a consists of eight data lines and one or more control lines. Control lines are needed to tell the receiving side when data is available (when the data lines are valid) and sometimes to acknowledge to the transmitting side that the data was received (a *handshake*). If this were a bidirectional interface, another control line indicating data direction would be needed, along with a mechanism to prevent both sides from transmitting at the same time.

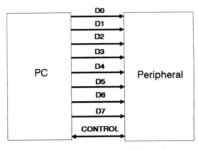

(a) 8-bit, unidirectional parallel interface

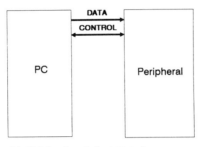

(b) Unidirectional, Serial Interface

Figure 8-1 Simple unidirectional digital interfaces: (a) parallel and (b) serial.

The serial interface in Figure 8-1b consists of only one data line (if it were bidirectional it probably would have two) and one or more control lines. In this scheme the data is time multiplexed. Control lines are used to indicate when the receiving end is ready to get the data along with other functions. The digital value of the data line represents a different bit at a different time. This requires a timing reference for the receiving end to decode the data accurately. When an external timing reference is used, this becomes a synchronous serial interface, with a control line carrying the required clock signal. When a receiver's internal timing reference is used, this becomes an asynchronous serial interface. To synchronize the incoming data stream with the internal clock, either a separate control line is used or, more commonly, a special start bit with a predetermined value is transmitted first. Then the data is sent, one bit per clock cycle, as shown in Figure 8-2.

Even though a parallel interface is inherently faster than an equivalent serial interface, it has drawbacks. Most parallel interfaces use standard digital logic voltage levels, usually TTL compatible. This limits their

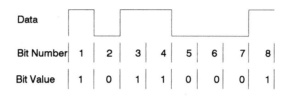

Figure 8-2 Sample eight bits of serial data.

noise immunity, where a long length of cable acts as an antenna, producing errors in the received data. In noisy environments, shielded cables are often required. In addition, long cables increase the capacitive coupling between adjacent signal lines, producing cross talk errors (a signal transition on one signal line induces a voltage spike in another signal line). Dispersion further distorts the signals as cable length increases. All in all, parallel interfaces have severe cable-length limitations, often on the order of just a few meters.

In contrast, most serial interfaces use much wider voltage swings to increase noise immunity (± 12 V is not unusual), and with few active signal wires, cross-talk noise is minimized. This enables serial interfaces to connect equipment hundreds of meters apart. Additionally, since fewer wires are required (and often shielding is not needed), serial interface cables are substantially less expensive (per foot) than parallel interface cables.

We will now explore some of these standard digital interfaces. First we will look at some common parallel interfaces. Later, we will examine several serial interfaces supported on PCs.

8.2 Parallel Interfaces

8.2.1 Centronics Printer Interface

The parallel printer interface, sometimes called the Centronics interface, is available for nearly all PCs and is supported by most printers. It is an 8-bit, unidirectional interface designed to transmit data from a computer to a printer, using TTL signal levels. The data usually sent is either ASCII codes, where each byte represents a printable character or a command (such as a Line Feed), or graphics data, consisting of command codes or data values (see Section 8.3.1 for a discussion of ASCII codes).

The standard IBM-compatible PC parallel printer port uses a 25-pin connector (DB-25) with the pin designations shown in Table 8-1. A special cable is used to connect this port to the 36-pin Centronics connector on most printers. The signal directions shown in Table 8-1 are relative to the PC. Signals with names starting with "-" (such as -ACK) are active low. The eight data lines, DATA0–DATA7, are unidirectional, sending data to the printer. The primary control and handshake lines in this interface are BUSY, -ACK, and -STROBE. BUSY goes low when the printer is ready to receive a new data byte. When the PC detects the printer is ready, it puts out data on the lines DATA0–DATA7 for a minimum of 500 nsec. Then it asserts the -STROBE signal for a minimum of 500 nsec, which tells the printer to read the data. The PC keeps the data lines valid for at least another 500 nsec.

In the meantime, the printer asserts BUSY and does its internal processing. When ready, it simultaneously negates BUSY and asserts -ACK. -ACK is asserted typically for 5 to 10 μsec. The -ACK line is virtually a redundant signal and usually the BUSY line alone is an adequate handshake for the PC, signaling data was received by the printer. The timing of this interface is shown in Figure 8-3.

The other parallel port control lines are used for various status and control functions. When -AUTO FEED XT is asserted by the PC, the printer automatically performs a line feed after it receives a carriage return. When the PC asserts -INIT for a minimum of 50 μsec, the printer is

TABLE 8-1
Parallel Printer Port Pin Assignments

PIN #	SIGNAL NAME	DIRECTION
1	-STROBE	OUT
2	DATA 0	OUT
3	DATA 1	OUT
4	DATA 2	OUT
5	DATA 3	OUT
6	DATA 4	OUT
7	DATA 5	OUT
8	DATA 6	OUT
9	DATA 7	OUT
10	-ACK	IN
11	BUSY	IN
12	PE	IN
13	SELECT	IN
14	-AUTO FD XT	OUT
15	-ERROR	IN
16	-INIT	OUT
17	-SELECT IN	OUT
18 - 25	GROUND	

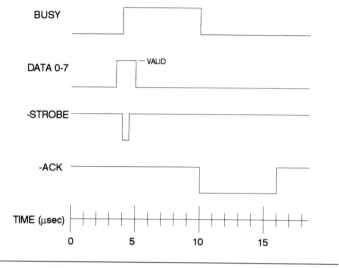

Figure 8-3 Parallel printer port interface timing.

reset to a known state (usually equivalent to its initial power-on conditions). When the PC asserts -SELECT IN, it enables the printer to receive data.

When the printer asserts PE it indicates it is out of paper. When the printer asserts SELECT it indicates it is enabled to receive data from the PC. When the printer asserts -ERROR it indicates that it is in an error state and cannot receive data.

An IBM-compatible PC can support up to three parallel printer ports designated LPT1, LPT2, and LPT3. Each port uses three consecutive I/O addresses. When a system boots up, DOS assigns the physical printer ports present to the logical LPT designations. LPT1 is assigned first, followed by LPT2 then LPT3. The starting addresses of parallel printer ports, in the order assigned to LPT designations, are 3BCh, 378h, and 278h. So, if all three ports are present in one system, port 3BCh becomes LPT1, port 378h becomes LPT2, and port 278h becomes LPT3. If port 3BCh is not present, port 378h becomes LPT1 and port 278h becomes LPT2. If only one parallel printer port is present, it is designated LPT1.

The printer port's starting address (3BCh, 378h, or 278h) is the data port, which can be an input or output. Writing to this port address latches eight bits of data on the DATA0–DATA7 lines sent to the printer. Reading from this port address returns the last byte latched (the real-time status of the output).

The printer port's next address (3BDh, 379h, or 279h) is the status port, which is read-only. It returns to the PC the value of the five status lines coming from the printer on the upper five bits of the port, as follows:

Bits 0–2 = unused
Bit 3 = −ERROR
Bit 4 = SELECT
Bit 5 = PE
Bit 6 = -ACK
Bit 7 = -BUSY

These lines can be polled for proper handshaking during a data output sequence. In addition, when -ACK is asserted (active low) it can generate IRQ7 (if enabled). This allows interrupt-driven software to handle printer output as a background task, for printer spooling. The printer would interrupt the PC, via its -ACK line, whenever it is ready to receive new data.

The printer port's next address (3BEh, 37Ah, or 27Ah) is the control port that can be an input or output. As an output, the PC latches the values of its control lines on the lower five bits of the port, as follows:

Bit 0 = -STROBE (1 = asserted)
Bit 1 = -AUTO FEED XT (1 = asserted)
Bit 2 = -INIT (0 = asserted)
Bit 3 = -SELECT IN (1 = asserted)
Bit 4 = IRQ EN (1 = asserted)
Bit 5 − 7 = unused

Note that most of the lines are inverted and asserted by a high bit except for -INIT, whose output follows the control port bit. The signal IRQ EN enables the port's IRQ7 output when bit 4 is latched high. As with the data port, a read from the control port will return the last value written to it.

The easiest way to use this parallel port to send data to a printer is with existing BIOS or DOS functions. Using the BIOS, INT 17h services the printer ports. It can either print a character (Function 0), initialize the printer (Function 1), or read the printer status (Function 2). On printing a character the proper handshaking protocol is used, with a time out if there is no response (if BUSY stays asserted indefinitely). The logical printer port (LPT) designation is used to select the desired printer. The BIOS does not support printer spooling, and special software must be used to support IRQ7 for printer output control.

A PC's parallel printer port can be used for other general purposes than printing, with certain limitations. It is ideal as a general-purpose output port with its eight unidirectional data lines, four output control lines, and five input control lines. Of course, there is no standard software support for using it this way, unless the standard printer interface handshake protocol (as in Figure 8-3) is adhered to. This would require special software to directly address the I/O ports used, to support a custom protocol.

The parallel printer port can also be used as a general purpose 5-bit input port, using the five status lines (-ACK, BUSY, PE, SELECT, and -ERROR). The real-time state of these lines can be read from the printer port's status register. In addition, the -ACK line can be used to generate IRQ7. The disadvantage here is having only five bits available for input and not being able to latch the data.

These limitations are addressed on IBM's PS/2 line of PCs. The parallel port on a PS/2 system has a fully bidirectional 8-bit data port, while keeping compatibility with the earlier implementation, as previously described. On this new parallel port, there is an extended mode that enables controlling the direction of the data port. Control port bit 5 (previously unused) now determines whether the data port is an output (bit 5 = 0) or an input (bit 5 = 1) port. The other control lines can now be used for different handshaking operations.

If you have a PC with an original IBM Printer Adapter Card, you can modify it to implement eight bidirectional data lines, analogous to the newer PS/2 parallel port. This should only be done if you have access to an accurate schematic for that card and are willing to run the risk of possibly damaging your hardware (and voiding your system's warranty). Of course, an IBM-compatible system may have very different circuitry.

Figure 8-4 shows part of an IBM printer adapter card. Writing to the data port stores output data in the 74LS374 octal latch, which goes to the DB-25 connector's data lines. Reading from the data port gives the real-time state of the data lines at the DB-25 connector via a 74LS244 buffer. The four output control lines (-STROBE, -AUTO FEED XT, -INIT, and -SELECT IN) are set in a 74LS174 hex latch by writing to the control port.

The output enable line (-OE) of the 74LS374 data output latch is normally grounded. This keeps the latch output permanently enabled. If the connection from line -OE (pin 1 of the 74LS374) to circuit ground is disconnected, the output of the latch can be disabled, allowing an external signal source to drive the DB-25 connector's DATA0–DATA7 lines. This external data could, in turn, be read from the data port (via the existing 74LS244 during a normal data port read cycle). The control for the octal latch's -OE line can come from any of the four interface control line. Line -INIT was chosen, since it goes to a logic 0 level when the board is reset, and a logic 0 level on this line enables the latch's output. To use the port

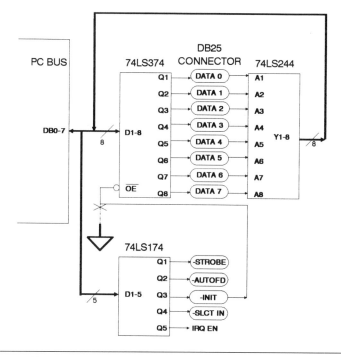

Figure 8-4 Modifications to a printer adapter card for 8-bit I/O.

for 8-bit input, writing a 1 to bit 2 of the control register turns off the output latch, allowing the reading of external data.

8.2.2 The IEEE-488 (GPIB) Interface

Another common parallel interface is IEEE-488 or GPIB (general-purpose interface bus). This interface is sometimes called the HPIB, as it was originally developed by Hewlett Packard to connect computers to their programmable instruments. GPIB was designed to connect multiple peripherals to a computer or other controlling device. Even though it was intended for automated instrumentation applications, it is also used to drive standard PC peripherals such as printers, plotters, and disk drives. It transfers data asynchronously via eight parallel data lines and several control and handshaking lines. All signals are at TTL voltage levels.

Instead of connecting one computer to one peripheral device, GPIB allows one computer to control up to 15 separate devices. In many ways, GPIB acts like a conventional computer bus. Each GPIB device has its own bus address, so it can be uniquely accessed. It uses a hardware handshaking protocol for communications, which supports slow devices. When communicating among fast devices, high data rates up to 1 Mbyte/second can be obtained.

The GPIB uses a master–slave protocol for data transfer. There can only be one bus master, or controller, at any given time. Typically, the master device is the controlling computer. A device on the bus has one of three possible attributes: Controller, Talker, or Listener. The controller manages the bus, sending out commands that enable or disable the talkers and listeners (usually, slave devices). Talkers place data on the bus when commanded to. Listeners accept data from the bus. A device can have multiple attributes, but only one at any given time. The computer can be a controller, talker, and listener; a read-only device, such as a plotter, will just be a listener; and a write-only device, such as a digital voltmeter, can be both a talker (when it reports a data reading) and a listener (when it is sent set-up information, such as a scale change).

The GPIB cable consists of 16 signal lines divided into three groups. The first group of signals consists of the eight bidirectional data lines, DIO1–DIO8. The second signal group consists of the three handshaking lines used to control data transfer: DAV, NRFD, and NDAC. The third signal group consists of five interface management lines that handle bus control and status information: ATN, IFC, REN, SRQ, and EOI.

The GPIB cable itself consists of 24 conductors, shielded, with the extra eight lines grounded. The cable is terminated with a special connector having both a plug and a receptacle, so that all the devices on the bus can be daisy-chained together, in either a linear or star configuration. Typically, the cable length between any two devices on the bus must be no more than 2 meters, while the total cable length of the entire bus must be no more than 20 meters. To exceed these limits, special bus extenders are needed. An additional limitation is that at least two-thirds of the devices on the bus must be powered on.

The GPIB uses standard TTL logic levels with negative logic, so when a control line is asserted it is at logic 0. This is because open-collector drivers are normally used on the bus interfaces. Therefore, a signal is pulled to a logic 1 level until a device asserts it and pulls it down to a logic 0 level. Figure 8-5a shows a simple GPIB linear configuration with four devices on the bus: a PC (controller), plotter (listener), meter (talker), and disk drive (listener and talker). Note that there is a separate cable connecting each pair of devices in the daisy chain. No special termination is needed for the last device.

Figure 8-5b shows schematically the electrical connection of a signal line (DAV in this example), with open-collector drivers drawn as a switch

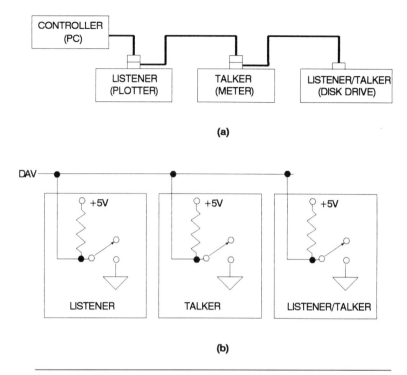

Figure 8-5 General-purpose interface bus (GPIB). (a) Typical GPIB linear configuration. (b) Open collector logic of a GPIB signal line (DAV).

to ground. Special line drivers specified for the GPIB are used on these interfaces to insure that when a device is not powered on it does not load down the signal line (the switch to ground is open). Even with special drivers, there is some leakage current to ground when a device is not powered on. That is why a maximum number of devices are allowed to be powered off when the GPIB is operational.

The pin designations for the standard GPIB connector is shown in Figure 8-6. As previously mentioned, the bidirectional data lines are signals DIO1–DIO8. The descriptions of the three handshake lines are as follows:

1. DAV (data valid) indicates when the data line values are valid and can be read.
2. NRFD (not ready for data) indicates whether or not a device is ready to accept a byte of data.

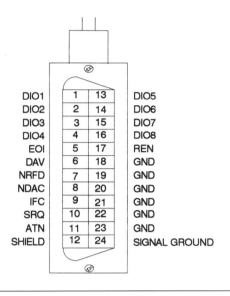

Figure 8-6 GPIB connector and pin designations.

   3. NDAC (not data accepted) indicates whether or not a device has accepted a byte of data.

The descriptions of the five interface management lines are as follows:

   1. ATN (attention) is asserted by the controller when it is sending a command over the data lines. When a talker sends data over the data lines, ATN is negated.
   2. IFC (interface clear) is asserted by the controller to initialize the bus when it wants to take control of it or recover from an error condition. This is especially useful when there are multiple controllers on a bus.
   3. REN (remote enable) is used by the controller to place a device in the local or remote mode, which determines whether or not it will respond to data sent to it.
   4. SRQ (service request) is used by any device on the bus to get the controller's attention, requesting some action.
   5. EOI (end or identify) is a dual-purpose line. It is used by a talker to indicate the end of the data message it is sending. It is also used by the controller requesting devices to respond to a parallel poll.

The sequence used to transfer data asynchronously on the bus, using the handshaking signals, is shown in Figure 8-7. This sequence is between an active talker (or the controller) and one or more active listeners. The speed of the transfer is determined by the slowest device on the bus. Initially, all the listeners indicate their readiness to accept data via the NRFD line. When a device is not ready, it pulls the NRFD line to a logic level 0 via its open collector output. As long as one active listener is not ready, NRFD is held low. Only when all active listeners are ready to receive data can NRFD go high (to logic level 1).

When the active talker (or controller) sees NRFD is high, it places its data byte on the bus (lines DIO1–8) and waits 2 μsec for the data bus to settle. Then it asserts DAV (to logic level 0), telling the active listeners to read the data. The listeners then pull NRFD low again, in response to the DAV.

The active listeners have all been holding NDAC active low. After DAV is asserted, as each active listener accepts the data on the bus it releases NDAC. When the last (slowest) listener releases NDAC, the signal goes high. The active talker (or controller) sees NDAC go high, negates DAV (goes high), and no longer drives the DIO lines.

Finally, the listeners recognize the negating of DAV and pull NDAC back low again, completing the transfer cycle. Now the handshake signals are ready for another data transfer to begin.

An important point is that this data transfer cycle is occurring between an active talker and one or more active listeners. Once the bus has been configured with talkers and listeners activated, the controller does

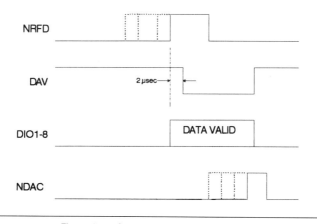

Figure 8-7   GPIB data transfer handshaking.

not have to be involved in the transfer (unless it is operating as a talker or listener). For example, a disk drive on the GPIB could send data to a printer on the bus without a computer's involvement, once the process was set up.

Two types of data are sent over the DIO lines: control data and message data. When the data flows from a talker to selected listeners, it is a message, which is machine-dependent data. This message data can either be an instruction for a device (e.g., change the output voltage on a programmable power supply) or data to/from a device (e.g., a voltage reading from a DMM). When a controller uses the data lines, it is sending control data (a command) to all the devices (both talkers and listeners) on the bus. The controller asserts the interface management line ATN to signal that this is a control data transfer (normally, it is negated for message data transfers). When ATN is asserted, any active talker releases the DAV line. The control data is sent by the controller using the same handshaking protocol described above. The major difference is that all devices on the bus receive this data, whether listener or talker and regardless of their active/inactive status.

The control data handles many aspects of the bus operation. It can configure devices as active listeners or talkers or it can trigger a device to perform its specific function. Each device of the bus has a unique 5-bit address (0–30). The controller can specify a device's address, enabling it as an active listener, for example, during a control data transfer cycle. Since control data commands are used for configuring the active talkers and listeners, it must be able to address all devices on the bus.

Device address 31 has a unique meaning for setting up listeners and talkers. If a control data command is sent to activate a listener at address 31, it actually deactivates all listeners. This is effectively the "unlisten" command. Similarly, when a control data command is sent to activate a talker at address 31, it deactivates the current talker. This is the "untalk" command. In addition, if a device is selected as the active talker, any talker that is currently active deactivates itself. This ensures that there is only one active talker at a time without requiring the bus overhead to explicitly deactivate the previous talker.

Another important GPIB management line is SRQ (service request), which is asserted by a device when it requires service from the controller. This may be an error condition in the device or an external event sensed by the device. Using SRQ is analogous to a processor interrupt, except that in this case the controller can ignore the SRQ or respond whenever it wants to. When the controller attempts to service the SRQ, it must first determine which device (or devices) is asserting the line. To do this it must poll all the devices on the bus.

There are two types of GPIB polling techniques: serial and parallel. In a serial poll, the controller issues a serial poll command, asserting ATN, to each device on the bus, getting back eight bits of status information. One of these status bits indicates whether the device issued the service request. The other bits convey device-dependent information. The main disadvantage with using a serial poll is that it is slow, requiring the controller to poll all the devices one at a time. Using a parallel poll is faster. In this case, the controller issues the appropriate parallel poll bus command, along with asserting the ATN and EOI lines. Up to eight devices on the GPIB can respond at once, setting or clearing the appropriate bit. In a parallel poll the only information obtained is which devices requested service.

So far, software aspects of the GPIB have not been mentioned, because they are not part of the IEEE-488 specification and can be device dependent. Every GPIB compatible device has its own unique set of commands. For example, a function generator would have a command telling it what type of waveform to output, and a programmable power supply would have a command for setting its current limit. These commands, and any appropriate responses such as the readings from a digital voltmeter, are all message data. Usually, message data on the GPIB consists of ASCII characters. The use of ASCII data for the GPIB is supported by HP, and the vast majority of GPIB equipment manufacturers.

Using a GPIB system can be very advantageous for complex data acquisition and control systems that require the high-level functionality of commercial test instruments. For example, consider a system required to characterize the frequency response of an electronic black box. Figure 8-8 shows a simple implementation using GPIB-compatible instruments: a function generator (to produce the variable excitation signal) and an AC voltmeter (to read the results).

A PC acts as the bus controller, using a commercially available GPIB interface card (see Chapter 11 for a sample of commercial sources). It controls the frequency and amplitude of the function generator's output (in this case a sine wave) and reads the AC voltmeter's input. Initially, the function generator should be directly connected to the AC voltmeter, to calibrate the system at its test frequencies. Then the device under test (DUT) is inserted between the generator and meter, and a new set of amplitude measurements is taken at the same set of frequencies. From this set of data, the transfer function or frequency response of the DUT can be calculated.

There is a large amount of software support for PC-based GPIB interfaces. Most GPIB interface cards for PCs come with software drivers for use with popular programming languages, such as C, BASIC, and

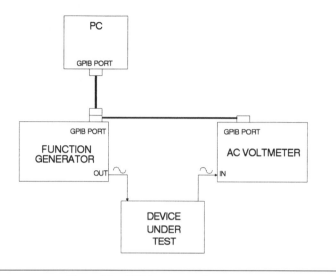

Figure 8-8 GPIB instrumentation example.

Pascal. Most high-end data acquisition software packages, such as ASYST or Labtech Notebook (see Chapter 11), support common GPIB cards, making the details of the GPIB operations invisible to the user. There are many other software packages with special features, making the process of implementing a GPIB system relatively painless. This is extremely useful due to the ever-growing number of instruments using the GPIB interface. GPIB equipment runs the gamut from power supplies and waveform synthesizers to digital storage oscilloscopes and network analyzers, to name just a few.

For example, National Instruments, a leading manufacturer of GPIB interfaces for a wide range of computers, provides a set of software drivers, called NI-488 software, for its MS-DOS PC-based products. The driver package is loaded into DOS using standard procedures. Then the special GPIB functions are called from the user's program. One of the languages supported by NI-488 is QuickBASIC, a compiled version of BASIC (see Chapter 13 for a discussion of programming languages). A simple program in QuickBASIC to take a reading from a digital multimeter is as follows:

```
CALL IBFIND("DMM",DMM%)
CALL IBWRT(DMM%, "FOROS2")
CALL IBRSP(DMM%,SPR%)
CALL IBRD(DMM%,DATA$)
PRINT DATA$
END
```

The first line in this program, calling IBFIND, retrieves initialization information on the specified device ("DMM") and returns the identifier code needed for the other functions. The second line, calling IBWRT, sends a device-specific message string to the DMM ("FOROS2"), configuring it for voltage type, range and speed. Next, the IBRSP call performs a serial poll on the DMM, checking its status. Finally, the IBRD call takes a voltage reading on the DMM and returns it in the string DATA$, which is then displayed by the print statement. In all of this, the user does not have to care about the details of the GPIB data transfers.

8.2.3 Other Parallel Interfaces

Before leaving the topic of parallel digital interfaces, it should be noted that there are many other standards besides the Centronics/printer interface and the GPIB. Most of these, such as BCD instrumentation interfaces or proprietary interfaces, have little or no support in the world of PC-based data acquisition equipment. One standard to be aware of here is the *small computer system interface*, or SCSI, which is usually used to connect high-speed disk drives to PCs. It is a general purpose, asynchronous parallel interface, originally eight bits wide, with newer implementations 16 bits wide. SCSI can be used to connect virtually any piece of equipment to a PC, including data acquisition devices. In practice, this is rarely done, except for older Macintosh computers that have only a SCSI interface instead of the newer NuBus.

8.3 Serial Interfaces

Many digital serial interfaces are in use. They are differentiated by many factors, including voltage levels, current drive capability, differential versus single-ended lines, single receiver and transmitter versus multi-drop capability, half versus full duplex, synchronous versus asynchronous, type of cable required, and communications protocols. These factors, in turn, determine important system specifications such as maximum data rate and maximum cable length. As we noted previously, the major reasons for using serial interfaces are low cable cost and potentially long cable lengths. The serial interfaces we will discuss here are all standards developed by the Electronic Industries Association (EIA) and are identified by their EIA standard number.

These standards define electrical characteristics and definitions of signal lines used in the interfaces. They do not define how the data will be sent or what each bit means. The two types of protocols used are asynchronous and synchronous. In an *asynchronous* protocol, the timing hardware at the transmitter and receiver are independent of each other (they are not synchronized). Synchronization is provided by the data stream itself, usually a particular level transition to indicate the start of data.

In a *synchronous* protocol, timing information is exchanged along with data, providing a single clock signal used by both ends of the interface. This allows serial transmissions at higher data rates than asynchronous protocols, since extra control bits indicating the beginning and end of a data byte are not needed, along with the extra time for an asynchronous receiver to synchronize itself to an incoming data stream. It is, however, a more complicated and expensive approach. Most PC-based serial data interfaces use an asynchronous protocol. We will discuss the commonly used asynchronous protocols in the following section, followed by a brief description of some common synchronous protocols at the end of this chapter.

8.3.1 The EIA RS-232C and RS-423A Interfaces

Without any question, the EIA RS-232C interface is the oldest and most common serial interface used by computer equipment. In fact, a PC's serial port is almost always RS-232C compatible. Due to its widespread use, RS-232C has paradoxically become one of the most nonstandard standards available. This is because it is used for much more than originally intended. RS-232C was developed in the 1960s as a standard for connecting data terminal equipment (DTE), such as the "dumb" terminals used with mainframe computers, to data communications equipment (DCE), such as a modem, over moderately short distances at modest data rates. Over the years, RS-232C evolved as a general-purpose interface between many varieties of equipment. One common example is connecting a personal computer to a printer or plotter. You can even use a special interface box to control a GPIB system via a computer's RS-232C port.

The RS-232C standard uses a 25-pin D-shell connector, with line designations as shown in Figure 8-9. Note that transmit and receive data directions are relative to the DTE end. RS-232C is a serial interface having two data lines to support full duplex operation. That is, the connected devices can simultaneously transmit and receive data, if they are capable. The maximum data rate on an RS-232C interface is 20,000 bits per second (bps) and the maximum cable length is 50 feet (although this can be increased at lower data rates or in low-noise environments). Note that EIA RS-232C can support either synchronous or asynchronous serial communications. In the vast majority of applications, asynchronous communications is used. However, the inclusion of two lines, Transmit Signal

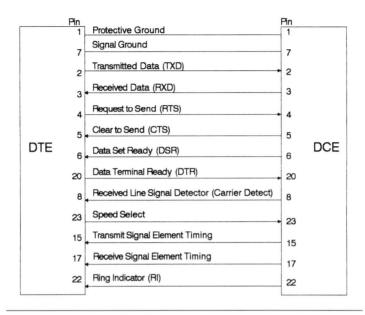

Figure 8-9 Standard RS-232C connections between data terminal equipment (DTE) and data communications equipment (DCE).

Element Timing and Receive Signal Element Timing, can provide the external clocking required by synchronous interfaces.

The RS-232C interface supports several handshaking lines, indicating each device's readiness to send or receive data. This is not an interlocking handshake, as used in GPIB for control of data flow. It simply enables or disables data transmission. These lines include Request to Send (RTS), Clear to Send (CTS), Data Set Ready (DSR), and Data Terminal Ready (DTR). The control lines, Ring Indicator (RI) and Received Line Signal Detector (or Carrier Detect, CD), are specifically used by modems.

On a PC, the usual RS-232C serial interface card supports asynchronous communications only and uses either a DB-25 or DB-9 connector. A PC/XT compatible system typically uses the 25-pin connector, with pin assignments as shown in Figure 8-10. Note that some of the EIA RS-232C standard signal lines are not used, such as those needed for synchronous communications. In addition, four non-RS-232C signals are added: +Transmit Current Loop Data, −Transmit Current Loop Data, +Receive Current Loop Data, and −Receive Current Loop Return. These lines support the 20-mA current loop interface used by older Teletype equipment.

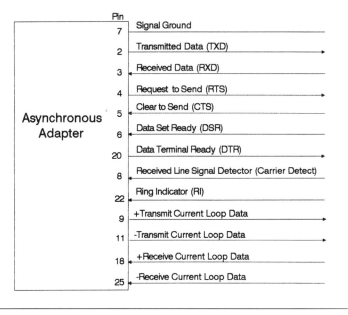

Figure 8-10 Pin designations for 25-pin asynchronous adapter.

An AT system usually has a 9-pin connector, with its pin assignments as shown in Figure 8-11. This limits the signals available to Transmitted Data (TXD), Received Data (RXD), DTR, DSR, RTS, CTS, RI, and CD. Usually a cable adapter is required to connect this 9-pin port to external devices with the conventional 25-pin D-shell connector.

Signals on RS-232C lines have well-defined electrical characteristics. Only one driver and one receiver are allowed on a line. The signals are all single-ended (unbalanced) and ground-referenced (the logic level on the line depends solely on that signal's voltage value relative to the signal ground line). The signals are bipolar with a minimum driver amplitude of ±5 V and a maximum of ±15 V (±12 V is the most common voltage used) into a receiver resistance of 3000 to 7000 ohms. Receiver sensitivity is ±3 V, so any signal amplitude less than 3 V (regardless of polarity) is undefined. Otherwise, a voltage level above +3 V is a logic 0 and below −3 V is a logic 1, as shown in Figure 8-12. Another important parameter is a maximum slew rate of 30 volts per microsecond. This means that an RS-232C signal running at the maximum voltage range of ±15 V must take at least 1 μsec to switch states.

If we look at the typical RS-232C application in Figure 8-13, where a terminal is connected to a modem, we see that most of the handshaking

8.3 Serial Interfaces 131

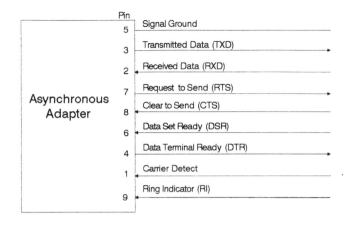

Figure 8-11 Pin designations for 9-pin asynchronous adapter.

lines act in pairs. When the terminal wants to establish communications, it asserts DTR. As long as the modem is powered on and operational, it asserts DSR as the handshake. These signals stay asserted as long as the communications link exists. When the terminal is ready to send data it asserts RTS. The modem generates a carrier signal on its analog line

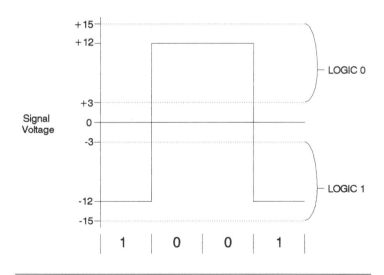

Figure 8-12 RS-232C signal levels.

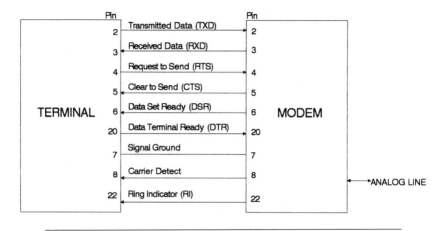

Figure 8-13 RS-232C connections between a terminal and a modem.

(usually a telephone line connection) and after a delay (allowing time for the modem on the other end to detect the carrier) it asserts CTS. Then the terminal can transmit its data over TXD.

When the terminal is finished transmitting, it negates RTS, causing the modem to turn off its carrier and negate CTS. If the modem now receives a carrier from a remote system over the analog line, it asserts CD. When it receives data from the remote system, it sends the data to the terminal over RXD. The cable used to connect the terminal to the modem is a straight-through variety. That is, pin 2 on one end goes to pin 2 on the other end, pin 3 on one end goes to pin 3 on the other end, and so on.

In actual practice, RS-232C interfaces are used to connect many different types of equipment. The asynchronous communications card in a PC (the serial port) is nearly always set up as a DTE (TXD is an output line and RXD is an input line—the opposite is true for a DCE device). The meaning of the handshaking line is software-dependent and may not have to be used. If required, just three lines can be used to minimize cable costs: TXD, RXD, and signal ground. If the software requires it, CTS and DSR must be asserted at the PC end for it to communicate, as when BIOS INT 14h functions are used for sending and receiving data over the serial port.

For example, if we want to send data between two PCs in the same room, without using two modems, we need a special cable, as shown in Figure 8-14. There are two approaches we can use to satisfy the handshake lines. In Figure 8-14a we implement full handshaking support, using

8.3 Serial Interfaces   133

seven wires. The data lines are crossed over, so TXD on one side is connected to RXD on the other side. Similarly RTS and CTS are crossed over as well as DTR and DSR. In this way, if the receiving end wants the transmitting end to wait, it negates the RTS line, which the other side sees as a negated CTS and CD and stops transmitting. Similarly, if one end wants to suspend communications entirely, it negates its DTR line, which the other side sees as a negated DSR. Signal ground is directly connected between the two ends. This cable, with the data and control lines crossed, is often referred to as a null modem cable. It is needed to connect a DTE to a DTE (or a DCE to a DCE).

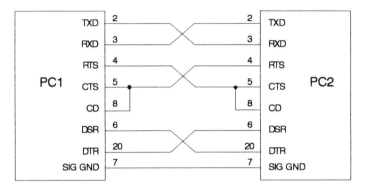

(a) Full Handshaking Support

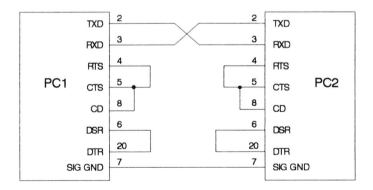

(b) No Handshaking Support

Figure 8-14   Connecting two PCs via an RS-232C cable.

A simpler connection, using only three wires, is shown in Figure 8-14b. In this case, the handshake lines are permanently enabled (*self-satisfying*) by connecting RTS to CTS and CD and connecting DTR to DSR at each PC. These lines cannot be used to control the data flow on the interface. The data flow can still be controlled, using special data characters in a software handshaking protocol. One popular software protocol widely supported is XON/XOFF. These are two ASCII control characters (XON is 11h, XOFF is 13h). When the receiving end needs to temporarily halt data flow, it sends an XOFF character to the transmitting end. When it is ready for data flow to resume, it sends an XON character. In a similar fashion, the ASCII characters ACK (06h) and NAK (15h) are also used for controlling data transmission. Using either of these software control protocols necessitates the use of ASCII data.

ASCII stands for the American Standard Code for Information Interchange. It is the most widely used computer code for handling alphanumeric (text) data and is usually employed for data transfers between equipment over standard interfaces. It is a 7-bit code consisting of printable (alphanumeric) and control characters, such as XOFF and CR (carriage return). The standard ASCII code is shown in Table 8-2. On IBM PC/XT/AT machines and compatibles, an eighth bit is added to the code producing special ASCII extension characters. These are nonalphanumeric printable characters, such as lines for character-based graphics.

TABLE 8-2
Standard ASCII Codes

b4	b3	b2	b1	0 0 0	0 0 1	0 1 0	0 1 1	1 0 0	1 0 1	1 1 0	1 1 1	
												b7 b6 b5
0	0	0	0	NUL	DLE	SP	0	@	P	`	p	
0	0	0	1	SOH	DC1	!	1	A	Q	a	q	
0	0	1	0	STX	DC2	"	2	B	R	b	r	
0	0	1	1	ETX	DC3	#	3	C	S	c	s	
0	1	0	0	EOT	DC4	$	4	D	T	d	t	
0	1	0	1	ENQ	NAK	%	5	E	U	e	u	
0	1	1	0	ACK	SYN	&	6	F	V	f	v	
0	1	1	1	BEL	ETB	'	7	G	W	g	w	
1	0	0	0	BS	CAN	(	8	H	X	h	x	
1	0	0	1	HT	EM	)	9	I	Y	i	y	
1	0	1	0	LF	SUB	*	:	J	Z	j	z	
1	0	1	1	VT	ESC	+	;	K	[	k	{	
1	1	0	0	FF	FS	,	<	L	\	l	|	
1	1	0	1	CR	GS	-	=	M	]	m	}	
1	1	1	0	SO	RS	.	>	N	^	n	~	
1	1	1	1	SI	US	/	?	O	_	o	DEL	

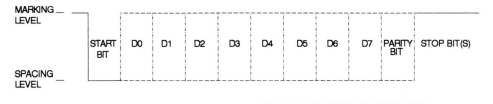

Figure 8-15 Asynchronous communications protocol.

As previously mentioned, the RS-232C standard does not specify the protocol used for data transmission. The vast majority of RS-232C interfaces, such as the IBM Asynchronous Communications Adapter and compatibles, use an asynchronous protocol. The transmission of one data byte using this protocol is shown in Figure 8-15. When no data is being transmitted, the line is at the *marking level*, which represents a logic 1. At the beginning of transmission, a start bit is sent, causing a line transition to the *spacing level*, a logic 0. This transition tells the receiver that data is coming. Next, the data bits (usually eight) are sent, one at a time, where a bit value of 1 is at the marking level and a bit value of 0 is at the spacing level. The data is followed by an optional *parity bit*, for error detection. Finally, one or more stop bits at the marking level are sent to indicate the end of the data byte. Since RS-232C line drivers and receivers are inverters, the marking level (logical 1) corresponds to a negative voltage (-3 V to -15 V), and the spacing level (logical 0) corresponds to a positive voltage ($+3$ V to $+15$ V) on the interface line.

The heart of an asynchronous communications adapter is the IC that converts parallel data to a serial format and serial data back to a parallel byte. This device is a Universal Asynchronous Receiver/Transmitter (UART). IBM and compatible asynchronous communications adapters use the National Semiconductor INS8250 UART IC in PC/XT machines and an INS16450 UART (which is a superset of the INS8250) in AT machines. This device has separate transmit and receive channels and control logic for simultaneously sending and receiving data. It produces its own programmable timing signals, from an on-board oscillator, for software control of data rates. It can send or receive serial data in the range of 50 bits per second (bps) to 38,400 bps. The width of each bit (in time) is the inverse of its data rate. So, at 9600 bps, each bit is $1/9600 = 0.104$ msec long. If 7-bit data is sent at this rate using a parity bit and only 1 stop bit (for a total of 10 bits per character), it would take 1.04 msec (0.104×10) to transmit a character. This would produce a maximum overall data transmission rate of 961 characters per second. This is not incredibly fast, but for small amounts of data it is acceptable. Bear in mind that many early serial terminals and modems ran at only 110 bps (which is nearly two orders of magnitude slower).

To set up an asynchronous RS-232C communications link, both machines (at the two ends of the line) must be set to the same data rate (sometimes incorrectly called the baud rate). In addition, the number of data bits must be known. It can often vary from 5 to 8 bits, although 7 or 8 bits are the most common. The next parameter needed is the parity bit. This is used as a simple error-detection scheme, to determine if a character was incorrectly received. The number of logical 1's in the transmitted character are totaled, including the parity bit. For even parity, the parity bit is chosen to make the number of 1's an even number, and for odd parity it is chosen to make the number of 1's odd. For example, the ASCII character "a" is 61h or 01100001 binary. For even parity, the parity bit would be 1 (making four 1's, an even number); and for odd parity, the parity bit would be 0 (leaving three 1's, an odd number).

When a parity bit is used (typically with 7-bit data characters), the transmitting end determines the correct parity bit value, as just described, and incorporates it in the character sent. The receiving end calculates the expected value of the parity bit from the character's data and compares it to the parity bit actually received. If these values are not the same, an error is assumed.

Of course, this scheme is not foolproof. It assumes that the most likely error will be a single wrong bit, which a parity check will always catch. If multiple bits are wrong in the same character, a parity error may not always be detected. Note that on IBM PC/XT/AT and compatible systems, the parity bit is not used with 8-bit data.

One final asynchronous communications parameter is the number of stop bits. This can be set to 1, 1-1/2, or 2 stop bits, although 1 bit is most commonly used. Unless very slow data rates are used, such as 110 bps, only 1 stop bit is adequate.

Several other single-ended serial communications interfaces are commonly used, besides RS-232C. One of these is RS-423A. This standard is sometimes used as an enhanced version of RS-232C, with several notable differences. RS-423A has a driver voltage output range of ± 3.6 V to ± 6.0 V, which is lower than RS-232C. However, RS-423A has much higher allowable data rates, up to 100K bps, and longer cable lengths (up to 4000 feet). One other notable difference is that RS-423A can support multiple receivers on the same line, up to a maximum of 10. This is very useful for unidirectional data transfers in a broadcast mode, such as updating multiple CRT displays with the same information. Table 8-3 shows the differences between several of the EIA transmission line standards.

8.3.2 The EIA RS-422A and RS-485 Interfaces

Another EIA serial transmission standard gaining increased popularity is the RS-422A interface, which uses differential data transmission on a balanced line. A differential signal requires two wires, one for noninverted data and the other for inverted data. It is transmitted over a balanced line, usually twisted-pair wire with a termination resistor at one end (the receiver side). As shown in Figure 8-16a, a driver IC converts normal logic levels to a differential signal pair for transmission. A receiver converts the differential signals back to logic levels. The received data is the difference between the noninverted data (A) and the inverted data ($\bar{A}$), as shown in the waveforms of Figure 8-16b. Note that no ground wire is required between the receiver and transmitter, since the two signal lines are referenced to each other. However, there is a maximum common-mode voltage (referenced to ground) range on either line of -0.25 V to $+6$ V, as shown in Table 8-3. This is because most RS-422A driver and receiver ICs are powered by the same $+5$-V power supply as other logic chips. Usually the signal ground is connected between the transmitter and receiver to keep the signals within this common-mode range.

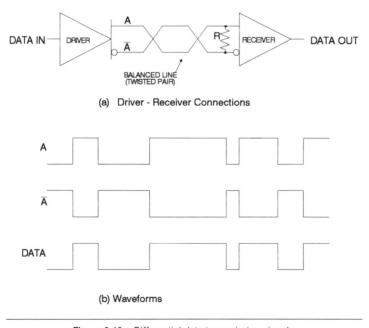

Figure 8-16 Differential data transmission signals.

TABLE 8-3
Comparison of Selected EIA Interface Standards

PARAMETER	RS-232C	RS-422A	RS-423A	RS-485
LINE MODE	Single-ended	Differential	Single-ended	Differential
MAXIMUM DRIVERS AND RECEIVERS	1 Driver 1 Receiver	1 Driver 10 Receivers	1 Driver 10 Receivers	32 Drivers 32 Receivers
MAXIMUM CABLE LENGTH	50 feet	4000 feet	4000 feet	4000 feet
MAXIMUM DATA RATE	20K bps	10M bps	100K bps	10M bps
MAXIMUM COMMON-MODE VOLTAGE	±25V	+6V -0.25V	±6V	+12V -7V
MINIMUM/MAXIMUM DRIVER OUTPUT	±5V min ±15V max	±2V min	±3.6V min ±6.0V max	±1.5V min
DRIVER OUTPUT RESISTANCE WITH POWER OFF	300 ohm	60K ohm	60K ohm	120K ohm
RECEIVER INPUT RESISTANCE	3K to 7K ohm	4K ohm	4K ohm	12K ohm
RECEIVER SENSITIVITY	±3V	±200 mV	±200 mV	±200 mV

This differential signal scheme enables the use of very high data rates (up to 10 M bps) over long cable lengths (up to 4000 feet) due to its high noise immunity. If external noise induces a signal on the transmission line, it will be the same on both conductors (A and $\bar{A}$). The receiver will cancel out this common-mode noise by taking the difference between the two lines, as shown in Figure 8-17. If a single-ended transmission line was used, the noise spikes could show up as false data at the receiver. The example in Figure 8-17 shows both a positive and a negative-going noise spike.

As with RS-423A, RS-422A can have multiple receivers (10 maximum) on the same line with a single transmitter. Again, this is basically useful for applications that require broadcasting data from a single source to multiple remote locations.

There are variations in the connectors and pin designations used for RS-422A interconnections. Most PC-compatible RS-422A interface cards use 9-pin D-shell connectors, but in lieu of an IBM standard, the pin designations employed vary from one manufacturer to another. An exam-

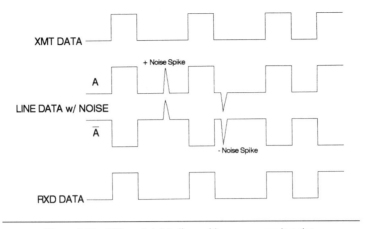

Figure 8-17 Differential data lines with common mode noise.

ple of the pin designations on a typical RS-422A interface card for PCs (from Qua Tech Inc.) is shown in Figure 8-18. Note that all the signal lines are differential.

The signal lines for AUXOUT are outputs and can be used to implement an RTS function. The signal lines for AUXIN are inputs and can be used to implement a CTS function. In this way, the RS-422A card can operate like a typical asynchronous RS-232C card in a PC (and use the same control software). Alternatively, the AUXOUT and AUXIN lines can be used to send transmit and receive clocks, for use with synchronous communications schemes.

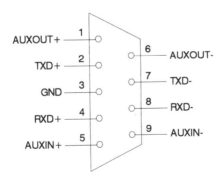

Figure 8-18 Pin designations for a typical RS-422A PC interface card.

The EIA RS-485 interface is basically a superset of the RS-422A standard. As shown in Table 8-3, its electrical specifications are similar to those of RS-422A. RS-485 is another differential transmission scheme, using balanced lines that can operate at speeds up to 10 Mbps over cable lengths up to 4000 feet long. It has somewhat different output voltage ranges, including a much wider common-mode range of -7 V to $+12$ V. The most important difference is that an RS-485 interface can support up to 32 drivers and 32 receivers on the same line. This allows actual networking applications on a party line system (sometimes called multi-drop) where all transmitters and receivers share the same wires.

To allow for this multi-drop capability, RS-485 drivers must be able to switch into a high-impedance (tri-state) mode, so that only one driver is transmitting data at any given time. As with RS-422A, all receivers can be active at the same time. A typical RS-485 multi-drop line is shown in Figure 8-19. Note that the termination resistor is typically placed at the last receiver on the line.

RS-485 interface cards for PCs are readily available and typically use the same connector (DB-9) and pin designations as similar RS-422A interface cards. The RS-485 driver output can be tri-stated using a control signal on the card. Usually a standard control line such as DTR is used for this, since it would not be used as an external line in a multi-drop interface. It is up to the software protocol to ensure that only one driver is enabled at any given time. One common way to do this is to use a master–slave relationship on the line. Only one driver/receiver station would be the master (or a network controller)—the others would be slaves. The master could transmit data at any time. The slaves could only transmit data after receiving an appropriate command from the master. Each slave

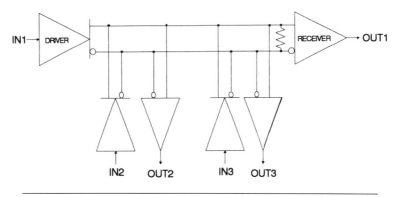

Figure 8-19 RS-485 multidrop application.

would have a unique ID or address on the line and would not be able to transmit unsolicited data. The high data rates available to an RS-485 network would compensate for the moderate amount of communications overhead required to implement a master–slave protocol and the constant polling performed by the master.

8.3.3 Synchronous Communications Protocols

As previously mentioned, synchronous serial communications protocols are much less common than their asynchronous counterparts in the world of personal computers, even though IBM does have synchronous communications adapters available for their PCs. Synchronous communication has noticeable advantages over asynchronous methods. Synchronizing bits (start and stop bits) are not needed, increasing the overall data transmission rate. Data does not have to be byte oriented (i.e., character-based) to be sent. In addition, it allows a system to communicate with large mainframe computers (especially IBM systems), which often use synchronous protocols. The drawbacks to using synchronous communications with PCs are higher costs for hardware and software along with limited support.

In synchronous transmissions, data is not always broken up into discrete characters, as with asynchronous methods. It tends to be block oriented, with a large amount of data (a block) transmitted at one time, with various control and error-checking information along with it. The data can be discrete characters (as with asynchronous methods) or bit oriented (no explicit data length). There are three common synchronous communications protocols: Binary Synchronous Communication (BSC), Synchronous Data Link Control (SDLC), and High-Level Data Link Control (HDLC).

BSC or *bisync* is a protocol developed by IBM. It is a character-oriented synchronous protocol where each character has a specific boundary. As with other synchronous protocols, there are no delays between adjacent characters in a block. Each block transmission may start with two or more PAD characters to ensure the clock at the receiving end of the line becomes synchronized with the clock at the transmitting end, even if a clock signal is being transmitted along with the data. Then, the start of the data stream is signaled by sending one or more SYN (synchronous idle) characters, which alerts the receiver to incoming data.

Next, one or more blocks of data are continuously sent. The data consists of characters 5 through 8 bits long with an optional parity bit, as with asynchronous methods. Often the data is encoded as ASCII characters, although it could also be EBCDIC (a code supported by IBM). Each block of data ends with an error-checking character, which provides much better data integrity than each character's parity bit. A popular error-checking technique used here (and in many other applications) is the cyclic redundancy check (CRC). The CRC takes the binary value of all the bits in the block of data and divides it by a particular constant. The remainder of this division is the CRC character, which will reflect multi-bit as well as single-bit errors.

IBM supports bisync on PCs with its Binary Synchronous Communications Adapter. This card uses an RS-232C compatible interface with a 25-pin D-shell connector. It is based on an Intel 8251A USART (Universal Synchronous/Asynchronous Receiver/Transmitter) IC. All the necessary protocol parameters are programmable, including mode of operation, clock source, and time out after no activity.

The other two popular synchronous protocols are SDLC and HDLC, which are both bit-oriented techniques where there are no character boundaries. The data is just a continuous stream of binary numbers, sent as an information field. This information field can vary from zero bytes up to the maximum allowed by the particular protocol in force. Like bisync, SDLC and HDLC data fields are framed by control information at the beginning and end. They also contain additional addressing information that makes them suitable for use with communications networks. HDLC contains more control information than SDLC. Unlike bisync, if transmission stops within an SDLC or HDLC field, an error is always assumed.

IBM supports SDLC for PCs with its Synchronous Data Link Control Communications Adapter. This card, as its BSC card, uses RS-232C compatible signal levels and a 25-pin D-shell connector. It is based on the Intel 8273 SDLC Protocol Controller IC.

This concludes our survey of common computer interfaces used by personal computers. In the next chapter we will look at data storage on the PC as well as data compression techniques.

CHAPTER

Data Storage and Compression Techniques

Acquired data must be permanently stored by the personal computer to allow future retrieval for display and analysis. The conventional storage devices available for personal computers use magnetic media (newer optical media storage devices are designed to look like magnetic media devices to the computer's software). Most of these storage devices (magnetic disk drives) use a random access, file-based structure. Magnetic tapes, for archiving (backup) applications, use a sequential structure.

Since nearly all application software, including data acquisition programs, assume data is stored on a magnetic disk (either a floppy diskette or a hard disk), these are the only storage devices we will consider here. Furthermore, since our main focus is on PC/XT/AT MS-DOS platforms, we will only consider DOS files in this discussion, although many of the basic principles covered will apply to other operating systems and non-80×86 computers.

9.1 DOS Disk Structure and Files

A file is a logical grouping of data, physically stored on a magnetic disk or other permanent storage media. The physical structure of a disk consists of concentric rings, called cylinders, and angular segments, called sectors, as shown in Figure 9-1. In addition, hard drives may consist of multiple platters (more than one physical disk in the drive package). The

144 CHAPTER 9 Data Storage/Compression Techniques

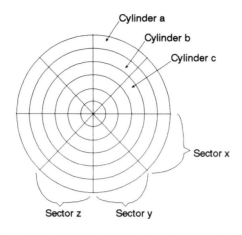

Figure 9-1 Physical organization of a magnetic disk surface.

cylinder on a single surface of a disk is referred to as a track. The read/write sensor used in a disk drive is the head. A double-sided floppy drive has two heads (one for each side of the diskette). A hard drive with four platters has eight heads. The read/write heads usually move together as one unit, so they are always on the same sector and cylinder (but not the same side of the platter or disk). Therefore, a physical location for data on a disk is specified by cylinder, sector, and head number.

The physical structuring of a disk into cylinders and sectors is produced by the DOS FORMAT program. In addition, FORMAT also initializes a disk's logical structure, which is unique to DOS. Each sector on every disk track (or cylinder) contains 512 bytes of data, along with header and trailer information to identify and delineate the data. This is why a formatted disk has lower storage capacity than an unformatted disk. The first sector (on the first cylinder) of every formatted DOS disk is called the boot sector. It contains the boot program (for a bootable disk) along with a table containing the disk's characteristics. The boot program, which is small enough to fit within a 512 byte sector, is loaded into memory and executed to begin running the operating system (DOS).

The boot sector is immediately followed by the file allocation table (FAT). The FAT contains a mapping of data *clusters* on the disk, where a cluster is composed of two to eight sectors. A cluster is the smallest logical storage area used by DOS. For floppy disks, a cluster is usually two sectors (1024 bytes); it is larger for hard drives. The FAT contains entries for all the logical clusters on a disk, indicating which are in use by a file and which are unusable (due to errors discovered during formatting). Each FAT entry is a code, indicating the status of that cluster. If the cluster is allocated to a file, its FAT entry points to the next cluster used by that file. The file is represented by a chain of clusters, each one's FAT entry pointing to the next cluster in the file. The last cluster in a file's chain is indicated by a special code in its FAT entry. This structure enables DOS to dynamically allocate disk clusters to files. The clusters making up a particular file do not have to be contiguous. An existing file can be expanded using unallocated clusters anywhere on the disk.

Due to the way file clusters are chained, a corrupted FAT will prevent accessing data properly from a disk. That is why DOS usually maintains a second FAT on a disk, immediately following the first one. This second FAT is used by data recovery programs (though not DOS itself) to "fix" a disk with a damaged primary FAT. Another side effect of the dynamic cluster allocation ability of DOS is that heavily used disks tend to become *fragmented*, where clusters for most files are physically spread out over the disk. This slows down file access, since the read/write heads must continuously move from track to track to get data from a single file. Several commercial utility programs are available to correct this, by moving data clusters on a disk to make them contiguous for each file and thus decrease file access time.

The FAT (and its copy) on a disk is followed by the *root directory*, which contains all the information needed to access a file present there. This information is the filename and extension, its size (number of bytes), a date and time stamp, its starting cluster number, and the file's attributes. The root directory is a fixed size (along with each file entry) so that DOS knows where the disk's data area, immediately following the root directory, begins. This limits the number of files that can be placed in the root directory. For example, a 360-Kbyte, double-sided floppy disk can only keep 112 entries in its root directory (which consists of four clusters of two sectors each). If more files must be stored on this disk, subdirectories have to be used. A subdirectory is a special file that contains directory information. It is available starting with DOS 2.0 and is used to organize groups of files on a disk. It is especially useful with large storage devices, such as hard drives.

Hard disks have one additional special area besides the boot sector, the FAT, and the root directory. It is called the partition table. This information describes how the hard disk is partitioned, from one to four logical drives. The table information includes whether a partition is bootable, where it starts, its ID code (it can be a non-DOS partition for another operating system, such as Xenix), and where it ends. To get around the disk size limitation of 32 Mbytes, in versions of DOS prior to 4.0 it is necessary to partition large hard disks into smaller logical drives. This is usually done with a special utility software package.

The directory structure of a DOS disk can be described as an inverted tree diagram, as illustrated in Figure 9-2. The root directory is symbolized by the backslash (\) character. The root has a limited number of possible entries, which can be either standard files or subdirectories. A subdirectory is a variable-size file (as are all DOS files), so its size and maximum number of entries is only limited by the free storage space available on the disk. Each subdirectory can contain conventional files along with other subdirectories. You can keep adding level after level of subdirectories. In Figure 9-2, the top level (Level 0) is the root directory, present on all DOS disks. Level 1 contains the first level of subdirectories (Sub1, Sub2, Sub3, Sub4), along with their files. Level 2 contains the subdirectories of Sub1 (Subsub1, Subsub2) and Sub3 (Subsub3, Subsub4). Level 3 contains the subdirectory of Subsub3 (Sss). Note that subdirectory names are limited to eight characters, as are all filenames. However, subdirectory names do not use a three-character extension, as other files do.

To access a file via DOS, the path to the directory containing that file must be specified, usually starting from the root (if the root isn't explicitly shown, the current default directory is assumed). In that path, directory levels are separated using the backslash (\) character. For example, \SUB1\SUBSUB2 would be the path to the SUBSUB2 directory. A \ character is also used to separate the directory path from the filename. For example, \SUB3\SUBSUB3\SSS\DATA.001 would be the complete file specification allowing DOS to locate the file DATA.001 in subdirectory SSS.

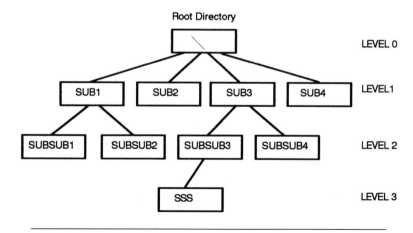

Figure 9-2 Example of DOS directory structure.

It should be noted that each directory level used on a disk requires DOS to search an additional subdirectory file to locate and access the file requested. If many directory levels are used (such as greater than five), DOS file access will be considerably slowed. You should use directories to organize your file storage logically, especially with a hard drive. Do not use more levels of subdirectories than you need.

For instance, you might have a hard disk subdirectory containing your data acquisition programs, called \ACQUISIT. You should keep your data files organized by projects or experiments, and separated into subdirectories, such as \ACQUISIT\PROJ1, \ACQUISIT\PROJ2, etc. However, there is no need to put each data file from the same project into its own subdirectory (\ACQUISIT\PROJ1\TEST1, \ACQUISIT\PROJ1\TEST2) unless they all have the same name. So, \ACQUISIT\PROJ1 may contain TEST1.DAT and TEST2.DAT.

9.2 Common DOS File Types

Standard DOS file types are denoted by a three-letter extension to the filename. We previously saw that .SYS files are loadable DOS drivers, for example. DOS files can be broken down into two broad categories: binary files and ASCII files.

In a binary file, data is stored in an unencoded binary format, just as it would appear in system memory. The end of a binary file is determined strictly from the file length recorded in its directory listing. Executable programs and device drivers are examples of the many types of standard binary files. Many data file formats are binary.

In an ASCII file, the data is stored as printable ASCII characters (see Chapter 8 for a discussion of the ASCII code). Each byte represents one ASCII character that is either printable or a special control character. The ASCII data is usually terminated by a control character signifying the end of the file. The file's directory listing still contains its file length. Various application programs, such as editors and word processors, typically operate on ASCII data files. We will now look at some of the standard DOS file types.

9.2.1 .BAT Files

Under DOS, filenames ending with the .BAT extension are considered *batch* files. A batch file contains DOS commands that will be automatically run, as if they were a program. Batch files have some rudimentary program capabilities, such as branching and conditional execution. For the most part, they are used to automate a group of commonly executed DOS commands, including calling application programs.

.BAT files are always ASCII files. They are usually created with an editor program, such as EDLIN (part of DOS). As an example, let us assume we want to copy all the files with a .DAT extension from a hard disk (drive C:) directory \TEMP to a floppy disk (drive A:) and then delete the original files. We can create a file named TRANSFER.BAT, with the following lines:

```
COPY C:\TEMP\*.DAT A:
DEL  C:\TEMP\*.DAT
```

These instructions will be carried out by DOS when we give the TRANSFER command (which executes TRANSFER.BAT). Note that a batch file is an interpreted program. DOS reads each ASCII line and then executes it. Therefore, it is relatively slow compared to performing the same function with a dedicated program.

A useful feature of DOS batch programs is the ability to employ variable data, which are ASCII strings. The contents of the variables used are specified at run time, when the batch file is executed. When the batch program is written, a percent sign (%) followed by a digit is used to represent the appropriate parameter supplied with the command to run the batch file (%1 is the first parameter, %2 is the second, and so forth). Using this feature, we can make TRANSFER.BAT more generalized, with the data files in \TEMP a variable:

```
COPY C:\TEMP\%1 A:
DEL  C:\TEMP\%1
```

To use this batch program to transfer all the .DOC files from C:\TEMP to A, use the command:

```
TRANSFER *.DOC
```

Batch files become more than just a list of commands when conditional statements are used. The following example is a file called HIDE.BAT, which changes a file's attribute to hidden, via the program SETATRIB. The variable parameter (%1) is the name of the file to hide:

```
ECHO OFF
IF EXIST %1 GOTO OK
ECHO "SYNTAX:   HIDE ⟨filename⟩"
GOTO END
:OK
SETATRIB HIDE %1
:END
ECHO ON
```

The ECHO OFF command tells DOS not to display the batch program lines as it executes them (normally it would). At the end of the program, ECHO ON turns this feature back on. The second program line checks to see if there was a parameter given with the batch file command, via IF EXIST. If there was, execution jumps to the label :OK, to execute the SETATRIB command. Otherwise, it displays the quoted text in the ECHO command (showing the proper syntax for the batch program) and jumps to the label :END, skipping the SETATRIB command.

One special batch file used by DOS is called AUTOEXEC.BAT. This file is executed by DOS after it boots up, if it exists in the root directory of the disk. It is used to perform many functions such as customizing system parameters (i.e., change the DOS prompt), calling an applications program needed at system startup (such as setting the DOS date and time from a real-time clock), or changing the default directory.

Batch files can handle fairly complex tasks but are best suited for simpler, commonly performed functions that do not warrant the time and trouble needed to develop a full-fledged program. The minimum functionality of the DOS batch facility also limits the tasks that can be performed by a batch file. In general, if you continuously repeat the same sequence of DOS commands, it is a good candidate for a batch file.

9.2.2 .BAS and Other ASCII Files

Many file extensions are commonly associated with ASCII files, although they are specified by application programs rather than by DOS itself. Some of these extensions are .TXT and .DOC. Even when ASCII is used by some applications, it is not always "plain vanilla" (exactly following the 7-bit ASCII code). Some word processing application programs mix ASCII with binary data in their files. Others use the eighth bit of each character for special text formatting commands (such as underlining), which ASCII does not directly support.

The DOS TYPE command displays an ASCII file on the video display. If the displayed text appears garbled or has nonalphanumeric characters (such as smiling faces), the file is not composed of plain 7-bit ASCII characters.

IBM BASIC and GW BASIC produce program files with the .BAS extension. These files are usually modified ASCII, using special characters called tokens to represent common BASIC commands. BASIC can save its program files in plain ASCII, if specifically instructed. BASIC also produces ASCII data files that can be used by a variety of application programs.

Many data acquisition and analysis programs will read or write ASCII data files. This is very useful, since the data can then be directly printed and easily reviewed by different people.

9.2.3 .COM Files

DOS files with the .COM extension are executable programs in a binary format. A .COM file contains a short program that must fit within a single 64-Kbyte memory segment, including all its data. The .COM file contains an absolute *memory image* of the program. The contents of the file are identical to the computer's memory contents when the program is loaded.

When the command to run a program is issued, either by the user at the DOS prompt or from another program via the DOS EXEC function call, DOS determines whether enough free memory exists to load the program. If not, it returns an error message. If there is adequate space, DOS determines the lowest available memory address to use. This memory area is called the program segment. At the beginning of the program segment (offset 0), DOS creates the program segment prefix (PSP) control block. The program itself is then loaded into memory at offset 100h of the program segment, since 256 bytes are reserved for the PSP. The PSP contains information needed to execute the program and return to DOS properly. After the program is loaded into memory, it begins execution.

A .COM program is automatically allocated all of the available system memory. If the .COM program wants to run another program without terminating itself first, via the DOS EXEC function call, it must first deallocate enough memory for this secondary program. Even though a .COM program must fit within a single 64-Kbyte memory segment, it can access memory outside of its segment by changing its segment pointers (such as the data segment pointer, DS).

Another ideosyncracy of .COM programs is that they must begin execution at offset 100h of their segment (immediately following the PSP). Since most .COM programs are written in Assembler, to minimize their size they would have the following statement just prior to the start of the program code:

$$\text{ORG} \quad 100\text{H}$$

This requirement is not a severe limitation, since the first program statement can be a jump to some other section of code in the segment.

9.2.4 .EXE Files

The second DOS format for executable programs is the .EXE file, which is another type of binary file. Programs in the .EXE format tend to be much larger and more complex than .COM programs. They can span multiple segments, both for code and data. In addition, they are relocatable and the exact location of various parts of the program are determined at execution time by DOS. Furthermore, they are not automatically allocated all available memory, as .COM programs are.

To accommodate this flexibility, .EXE files begin with a special header area. The first two bytes of this header begin with 4Dh and 5Ah (in ASCII, "MZ") to indicate to DOS that this is an .EXE program. The rest of the header contains various information including the length of the program, the length of the file, its memory requirement, relocation parameters, and where to begin program execution. Unlike .COM files, .EXE programs do not have a fixed starting point for program execution. In an .EXE file, the header is immediately followed by the program itself.

When DOS attempts to run an .EXE program, it first reads the header, determines whether enough free memory is available, creates the PSP, loads the program, and starts its execution. Because of their larger size and the extra work DOS must do, .EXE programs tend to load more slowly than .COM programs. The vast majority of commercial applications are .EXE programs. Some are so large that they need more than the maximum available DOS memory area of 640 K. They typically make use of overlays to accommodate large code areas and use expanded memory (when available) to handle large data-area requirements.

When a program is developed using a standard compiler (such as Macro Assembler, C, Pascal or FORTRAN) under DOS, an .EXE file will be produced by the final linking process (see Chapter 13 for a discussion of programming languages and the various compiling processes). If the program was written to fit within a single 64-Kbyte segment, it can be successfully converted into a .COM file, using the DOS program EXE2BIN. If program file size or load time do not need to be minimized, it is not necessary to convert an .EXE program into a .COM program. When given the choice between the two executable program formats, it is usually advantageous to keep the flexibility of an .EXE program.

9.3 Data Compression Techniques

Data acquisition applications usually involve the creation and storage of large amounts of unprocessed data. If a particular test was acquiring 8-bit data at the modest rate of 5000 samples/second, one minute of data would require 300,000 bytes of storage. Ten minutes of unprocessed data would require nearly 3 Mbytes of storage. Data at this rate could fill a hard drive after a relatively small number of tests. That is why data compression techniques are so important.

If large amounts of data need to be transferred between remote systems, data compression not only reduces the storage requirements for the data—it also reduces the transfer time needed (and its inherent cost). If data is being sent serially via modem, even at the relatively fast rate of 9600 bps it would take over 2 minutes to transfer 100 Kbytes of data.

Many different techniques are employed to reduce the storage requirements of large amounts of data. The most important measurement of a particular technique is its compression ratio: the size of the original data divided by the size of the compressed data. Another important parameter of a data compression technique is its fidelity or distortion. This is a measure of the difference between the original data and the compressed/restored data. In many applications, no data distortion can be tolerated, such as when the data represents a program file or an ASCII document. A relatively low compression ratio would be expected then. In other cases, a small but finite amount of distortion may be acceptable, accompanied by a higher compression ratio. For example, if the data in question represents a waveform acquired at a relatively high sampling rate, storing every other point is equivalent to filtering the waveform and producing a small amount of distortion, particularly for high-frequency components in the data.

Thus, the nature of the data dictates the parameters important to the data compression process and helps indicate which technique is best suited. The general tradeoffs are between compression ratio and fidelity. An additional factor, usually less important, is the amount of time required to compress or restore the data, using a particular technique. This can become an important factor if the data compression is done in real time, along with the data acquisition or transmission.

We will now look at various data compression techniques and their appropriate applications. Most of the techniques, unless otherwise noted, are primarily useful for files containing numerical data.

9.3.1   ASCII-to-Binary Conversion

Sometimes there are very obvious ways to reduce the size of a data file. If a set of numerical data is stored in an ASCII format (as many commercial data acquisition application programs are), encoding it directly as binary numbers could produce large space savings. For example, if the data values are signed integers within the range of $\pm 32{,}767$, they can be represented by two bytes (16 bits) of binary data. These two bytes would replace up to seven ASCII character, composed of up to five digits, one sign character, and at least one delimiter character separating values. This ASCII-to-binary conversion would produce a maximum compression ratio of 3.5:1 with no distortion. Even if the average value used four ASCII digits (1000–9999) the compression ratio would still be 3:1. After this conversion, other techniques could also be applied to the data set, further increasing the data compression.

9.3.2 Bit Resolution and Sampling Reduction

When a set of data represents numerical values, as in a waveform or data table, the number of bits used to represent those values determines the minimum resolution and the maximum dynamic range. As we saw previously, the minimum resolution is the smallest difference that can be detected between two values, which is one least significant bit (LSB) for digitized numbers. The ratio between the maximum and minimum measurable values determines the dynamic range:

$$\text{dynamic range (in dB)} = 20 * \log(\max/\min)$$

If lowering the resolution can be tolerated, data compression can be easily and quickly implemented. The resulting compression ratio is simply the original number of bits of data resolution divided by the new (lower) number of bits.

As an example, let us assume we have a set of data acquired by a 12-bit ADC system, with a dynamic range of 4096:1 or 72 dB. We first search the data set for the minimum and maximum values (we will assume the data is represented as unsigned integer values, for simplicity). In this example, the minimum value is 17 and the maximum is 483. A data range of 17–483 can be represented by nine bits without any loss of resolution (or fidelity) for a compression ratio of 12/9 = 1.33:1. If the minimum value was larger, such as 250, the difference between maximum and minimum, now 233, can be represented as fewer bits (8, for a range of 0–255) than the range of 0 to maximum value (483). In this case, we can get a compression ratio of nearly 12/8 = 1.5:1 by subtracting the minimum value from all the data points. The minimum value must then be included with the 8-bit data, so the correct values can be reconstructed. Adding a single 12-bit value to the compressed data is very little overhead when many points are contained in the waveform.

The simple technique above is useful when the acquired data does not fill the entire dynamic range of the data acquisition system. Then, the unused bits of resolution can be discarded without causing any data distortion. Most of the time, we do not have this luxury. To compress a set of data we usually have to sacrifice some resolution.

Still using a 12-bit data acquisition system, let us assume a data set has a minimum value of 83 and a maximum value of 3548. Now, minimum − maximum = 3465, which still requires 12 bits of resolution. If we have to compress this data, we will lose some resolution. Assuming we need a minimum compression ratio of 1.5 : 1, we can normalize the data to 8 bits. To do this, we multiply all the data values by the new maximum value (255 for 8 bits) and divide them all by the original maximum value (3548). The number 3548/255 = 13.9 is the scaling factor. Either this scaling factor or the original maximum value is kept with the normalized data, to enable its restoration to the proper values and dynamic range. The data can be restored to its full dynamic range, but its resolution will be 14 times coarser, due to the rounding-off that occurred when the data was normalized. Any two original data points that were separated by values of less than 14 will no longer be distinguishable. So, if two data points had original values of 126 and 131, after normalizing to 8 bits (dividing by 13.9), they will both be encoded as 9 and restored as 125.

Figure 9-3a shows a simplified flowchart for an algorithm that compresses a set of data by reducing the number of bits used to represent it. As we see, this approach can produce a loss of fine details, due to lower resolution. To exploit this form of compression, the data must be stored efficiently. Figure 9-3b shows data compressed to 6 bits per value. Four point values are stored in three data bytes, where each byte contains the bits from two adjacent values.

Another simple approach, often more acceptable than extreme resolution reduction, is sampling reduction. If the maximum frequency content of the digitized data is well below the Nyquist frequency, the effective sampling frequency can be reduced. For example, if an original set of data was filtered to limit its high-end to 1 kHz while being sampled at 10 kHz, the Nyquist frequency is 5 kHz. If every pair of adjacent values was averaged and stored, the effective sampling rate would be reduced to 5 kHz and a compression ratio of 2 : 1 would result. For this new set of data, the Nyquist frequency is also reduced by 2 to 2.5 kHz, still well above the maximum frequency content of the data.

This sample compression technique still distorts the data, as does the bit compression previously described. Still, if the high-frequency data artifacts lost are mostly noise, there is no harm done.

9.3.3 Delta Encoding

Another popular technique for compressing strictly numerical data is *delta encoding*. This approach is especially useful when the data represents a continuous waveform with relatively low instantaneous slopes. In such a set of data, the difference between adjacent points is small and can be represented by far fewer bits than the data values themselves. Delta

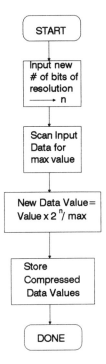

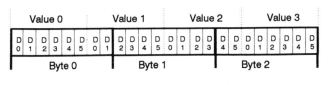

(a) Simplified Flow Chart for Resolution Reduction Algorithm

(b) Resulting data packing with n = 6 bits per value

Figure 9-3 Data compression via resolution reduction.

encoding consists of keeping the first value of the data set, at its full bit resolution, as the starting point. All subsequent values are differences, or deltas, from the previous value, using fewer bits.

To illustrate this, Table 9-1 contains a data set of 11 original values, which require 12 bits each for full binary representation. The delta-encoded values start with the first original 12-bit value. The next value is +20, the difference between the second and first values. The next delta-

TABLE 9-1
Example of Delta Encoding a Small Data Array

Original Values	Delta Encoded Values
3125	3125
3145	+20
3175	+30
3185	+10
3193	+8
3201	+8
3192	-9
3183	-9
3170	-13
3152	-18
3130	-22

encoded value is +30, the difference between the third and second values. This continues until the delta between the last and next to last values is computed. Examining the delta-encoded values shows us that they all fit within the range of ±32 and can be represented by six bits (one bit is for the sign). If we do use six bits for each delta value, the delta-encoded data set would require 10 * 6 + 12 = 72 bits for storage (remember, the first value is at full 12-bit resolution), compared to 11 * 12 = 132 bits for the original data set. The compression ratio here is 1.83:1. It will approach 2:1 as the size of the data set grows and the overhead of the first 12-bit value becomes negligible.

The key to getting high compression ratios with delta encoding is to use as few bits as possible to represent the delta values. One common problem is that with most data sets, a small number of bits can represent most of the delta values, while a few deltas require many more bits, due to occasionally high local slopes or transient spikes. Instead of increasing the number of bits for delta representation to accommodate a very small number of anomalous values, an escape code can be used. Let us assume that our data set is still using a 6-bit delta representation (±31), and a delta value of +43 comes along. We can designate one of the least-used delta values as the escape code—either $+31$ or -31 would be a good choice. This escape code would be followed by the full-resolution 12-bit value, which cannot be represented by a small delta value. After this number, delta values continue as before. So, if we had a data set with 128 12-bit numbers, using 6-bit delta encoding that handled all but three values, the total number of bits encoded would be

$$127 * 6 + 4 * 12 = 810$$

for a compression ratio of 1.9 : 1. If the three anomalous values could be accommodated by 8-bit delta values and no escape codes were used, the total number of bits would be

$$127 * 8 + 12 = 1028$$

for a compression ratio of 1.5 : 1. Obviously, the judicious use of escape codes for the infrequently large delta value will produce the best compression ratio. If the escape code is used too often, the compression ratio can decrease severely (it could even become less than 1 : 1 if a large fraction of values use the escape code).

With the appropriate data set, delta encoding can produce reasonable compression ratios with no data distortion. If it is combined with a statistical technique, such as Huffman encoding (described below), even higher compression ratios can be obtained, without any data distortion. One drawback to delta encoding, especially when used to transfer data via potentially error-prone means (such as over telephone lines via modems) is that once an error occurs in the compressed data set, all values following it will be erroneous. As with any other set of compressed or encoded data, it is always a good idea to include error detection information with the data, such as a checksum or CRC. If the block of data is large enough (such as several hundred bytes) the overhead from the few extra error detection bytes will have a negligible impact on the overall compression ratio while increasing the integrity of the data tremendously.

9.3.4 Huffman Encoding

Many compression techniques are based on statistical relationships among items in a data set. One of the more popular statistical methods is *Huffman encoding*. This technique will only work if a relatively small number of data set members (possible numerical values or characters) have a high probability of occurrence. If nearly all possible values (or characters) have equal probability of occurrence (a random distribution), this method will produce a compression ratio of less than 1 : 1.

Basically, Huffman encoding employs a variable number of bits to represent all possible members of the data set. Data set members with a high probability of occurrence use the smallest number of bits (less than the unencoded number of bits), while those members with very low probabilities use larger number of bits (sometimes larger than the unencoded number of bits). The bit values are chosen so that there is no confusion in decoding the encoded data. Huffman encoding produces no data distortion. The restored data is identical to the original, uncompressed data. The amount of data compression produced by this technique varies with the statistical distribution of the data set used.

ASCII data representing English text is commonly compressed using Huffman encoding, since the probability of occurrence of the various alphanumeric characters is well known. Certain vowels, such as e or a, or even the space character, will occur very frequently while other characters, such as x or z, will occur very rarely. The common characters may need only three or four bits to represent them in a Huffman code, while the uncommon ones may require more than seven or eight bits. A typical ASCII document may average around 5 bits per character using Huffman encoding. If the original data was stored as 8-bit characters, this produces an average compression ratio of 1.6:1.

Huffman encoding is often used with other techniques, such as delta encoding, to further increase a data set's compression ratio. To implement Huffman encoding, the statistical probability of occurrence of each possible data set member (numerical value or ASCII character, for example) must be known. Table 9-2 shows a simple example for a set of 4-bit delta-encoded values, in the range ±7. Only a few delta values have very high probabilities. Just five of the possible 15 delta values account for 80% of the data set (±1 = 20%, ±2 = 15%, 0 = 10%). In fact, a crude figure of merit can be calculated by taking this major subset of data values and dividing its total probability of occurrence (here, 80%) by the fraction of possible values it represents (in this case 5/15 = 0.33). For our example, this figure of merit is 0.80/0.33 = 2.4, which is good enough to warrant using Huffman encoding. A figure of merit below 2.0 would not be very promising for Huffman encoding.

Figure 9-4 shows a graphical method used to implement Huffman encoding. This approach is only manageable with small data sets, as in our example. The algorithm can readily be translated into a computer program for data sets with a large number of members (such as 7-bit ASCII characters).

First, we start with the data set of 15 possible values, listed in order of probability of occurrence, from Table 9-2. The data values (deltas) are listed across the top of the figure along with their probabilities (in paren-

TABLE 9-2
Data for Huffman Encoding Example

DELTA VALUE	PROBABILITY	HUFFMAN CODE	# OF BITS
+1	0.20	00	2
-1	0.20	01	2
+2	0.15	100	3
-2	0.15	110	3
0	0.10	1010	4
+3	0.05	1110	4
-3	0.05	10110	5
-4	0.02	11110	5
-5	0.02	101110	6
+4	0.01	1011110	7
+5	0.01	1011111	7
+6	0.01	1111100	7
+7	0.01	1111101	7
-6	0.01	1111110	7
-7	0.01	1111111	7

theses), which should all add up to 1.00. To start, we draw pairs of lines connecting the lowest probability values—in this case, the .01 values at the left side of the diagram. At the vertex of the two lines connecting these pairs, we write the sum of their probabilities (.02, in this case). We continue pairing off and summing probability values until all the values are used and the overall sum at the bottom of the diagram is 1.00.

Now, we arbitrarily assign a binary 1 to every line that points up to the left and a binary 0 to each line that points up to the right, differentiating the paths used to get from the 1.00 probability value up to the original delta value. Finally, each line connecting the final 1.00 vertex to a delta value's starting point, at the top, represents a bit. We could have just as easily reversed the 1's and 0's. The code for each delta value is the concatenation of bit codes used to trace its path, starting at 1.00.

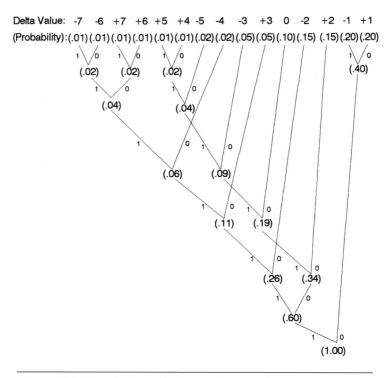

Figure 9-4 Example of graphical approach to determining Huffman codes.

So, the Huffman code for +1 is 00 and the code for −1 is 01, each only 2 bits long. The paths for +2 and −2 use three lines (for three bits) and are, respectively, 100 and 110. All the other delta values are assigned their codes in the same way. Values 0 and +3 use four bits, −3 and −4 use five bits, −5 uses six bits, and all the other values use seven bits. As we see, the delta values with the highest probabilities use the smallest number of bit.

When the encoded data is restored, the codes with the smallest number of bits are tested first. If no match is found, the number of bits tested expands until a valid code is located. If no valid code is determined after examining the maximum number of bits, an error is assumed.

If we use the Huffman codes in Table 9-2, let us see how the following encoded binary string would be decoded:

111000101111001

First, we look at the first two bits, 11, which are not a valid 2-bit code (only 00 or 01 are valid). Looking next at the first three bits, 111, we do not see a valid 3-bit code (only 100 and 110 are valid). When we check the first four bits, 1110, we find a valid code for +3. The remaining bits are now:

$$00101111001$$

The first two bits here, 00, are a valid code for +1. We are now left with:

$$101111001$$

Here, there are no valid 2, 3, 4, 5, or 6-bit codes. The first code to match is the 7-bit code for +4, 1011110. The remaining two bits, 01, are the valid code for −1. So, the decoded delta in this 15-bit binary string are +3, +1, +4, and −1. Of course, in a practical implementation, a program would use this search algorithm.

We can calculate the average number of bits a delta entry from Table 9-2 would use when encoded this way, and hence, the compression ratio. We just sum the product of the probability times the number of bits in the Huffman code for each delta value:

$$m = p_0 * n_0 + p_1 * n_1 + \cdots + p_k * n_k$$

where

 m = average number of encoded bits
 p_i = probability of occurrence for the ith data set value
 n_i = number of encoded bits for ith data set value
 k = number of values in the data set

If n is the number of bits per value in the original data set, the compression ratio is simply n/m. In our example, $m = 3.19$ bits and the compression ratio is $4/3.19 = 1.25:1$, which is not very large. However, since the data were already delta encoded, the original compression ratio (say, $2:1$) is multiplied by the Huffman encoding compression ratio ($1.25:1$) to give a larger overall compression ratio ($2.5:1$). Sometimes, this particular combination of compression techniques is referred to as delta Huffman encoding.

If a data set contained many more members than this previous example while maintaining a large percentage of values represented by very few members (with a large figure of merit), the compression ratio provided by Huffman encoding would be much larger. As with delta encoding, it may be useful to implement an escape code for the rare value that will not fit within the set of encoded values. In our example, it would be a delta value greater than +7 or less than −7. By its nature, the escape code would be a very low probability code, with a relatively large number of bits.

9.3.5 Significant Point Extraction

Some compression techniques are used exclusively on data points that constitute a waveform. *Significant point extraction* is a generalized technique that reduces the number of points required to describe a waveform. This approach causes varying degrees of data distortion, but can provide large compression ratios (in the range of 5:1 to 10:1, for example).

Significant point extraction operates on a digitized waveform, consisting of either a one-dimensional array of amplitude (y) values acquired at known, constant time intervals or a two-dimensional array of (x,y) coordinates. The one-dimensional array is the most common form of storage for values saved by a data acquisition system. The data is analyzed point by point to see where a group of adjacent points can be replaced by a straight line. The discarded point values can be estimated by interpolating from this line. Only the significant points required to produce a close approximation of the original waveform are retained.

Figure 9-5a illustrates a typical digitized waveform with significant points indicated by X characters. If the original waveform was composed

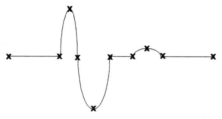

(a) Original Waveform with Significant Points noted by x

(b) Waveform Reconstructed from 10 Significant Points

Figure 9-5 Example of significant point extraction and reconstruction.

of 100 points, extracting only 10 significant points produces a 10:1 point compression ratio (the actual byte compression ratio will be smaller). The significant points include the waveform boundary points (start and stop) as well as places where the slope and/or amplitude change dramatically. Figure 9-5b shows the waveform reconstructed from the significant points only. Note that some of the finer details are lost, while the gross waveform structures remain. The acceptability of this distortion depends on the application of the waveform data. Often, the distortion is determined quantitatively, such as by the RMS (root mean square) deviation of the reconstructed data points from the original data points:

$$d = \{[(n_1 - m_1)^2 + (n_2 - m_2)^2 + \cdots + (n_j - m_j)^2]/j\}^{1/2}$$

where

d = RMS distortion
n_i = value of ith original point
m_i = value of ith restored point
j = number of points in waveform

One method of determining the significance of a point in a waveform is to calculate its *local curvature*. This is a measure of how much a waveform deviates from a straight line in the vicinity of a point. To illustrate, Figure 9-6a contains a simple waveform with one peak, composed of 23 points. To calculate local curvature, we pick a window size—in this case ±3 points—to consider the curvature around each point. If this window is too small, the calculation is not very significant. If the window is too large, local details tend to be averaged and lost ("washed out"). If the window is $2n$ points wide, we first start looking n points from the end of the waveform, in this case from the left side.

Since this is a one-dimensional array, the x-direction increment is constant for each point and we only need to look at data in the y direction (amplitude). For each point i, we do two scans from left to right: The first scan starts at point $i - n$ and ends at point i and the second starts at point i and ends at point $i + n$. This mens that we cannot scan the first or last n points in the waveform completely. For the first scan, we have two counters: $dy+$ and $dy-$. Starting with the leftmost point in the scan window, if the next point is more positive than the previous point we increment $dy+$; if it is more negative, we increment $dy-$ (if it is unchanged we leave the counters alone). We continue with the next pair of points until we get to the end of our scan (point i). The second scan starting at point i is similar, except now if the new (rightmost) point is more positive than the previous point we decrement counter $dy+$, and if it is more negative we decrement $dy-$ (if it is unchanged we leave the counters alone). After

164 CHAPTER 9 Data Storage/Compression Techniques

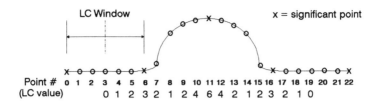

(a) Measuring Local Curvature (LC) with 3 Point Wide Window

(b) Reconstructed Waveform from 5 Significant Points

Figure 9-6 Using local curvature maximas to determine significant points.

completing the $\pm n$ points scan, the local curvature (lc) is the sum of the absolute values of these two counters:

$$lc = |dy+| + |dy-|$$

In Figure 9-6a, we cannot calculate the local curvature for points 0–2 and 20–22. Starting at point 3, after the first scan (from point 0 to point 3), $dy+ = 0$ and $dy- = 0$. After the second scan (from point 3 to point 6), $dy+ = 0$ and $dy- = 0$. So, for point 3, the local curvature is 0, or $lc(3) = 0$. For point 4, from the first scan $dy+ = dy- = 0$, while from the second scan $dy+ = 1$ (since point 7 is greater than point 6) and $dy- = 0$. So, $lc(4) = 1$. These calculations of lc continue for the rest of the waveform, up to point 19. We notice at the peak, $lc(11) = 6$.

Once the lc values are calculated, we can pick the significant points as the locations of the local curvature maximums. In this example, these are points 6 ($lc = 3$), 11 ($lc = 6$), and 16 ($lc = 3$). We also keep the first and last points (0 and 22) of the waveform as significant, since they are the boundaries. Therefore, we have reduced a 23-point waveform to five points, for a point compression ratio of 4.6:1. Figure 9-6b shows the waveform reconstructed from these five significant points.

There are many variants on using this local curvature technique to extract significant points. A minimum threshold could be selected that maximum lc values must reach before the corresponding point is considered significant. Another approach is to use amplitude weighting in the lc calculations. The $dy+$ and $dy-$ counters, previously described, produce an unweighted measure of local curvature, where a large-amplitude change counts as much as a small change in the same direction. They could be weighted by the relative amount of amplitude change, not just direction. When $dy+$ and $dy-$ would ordinarily be incremented or decremented by 1, they now increase or decrease by the amount of amplitude change between two adjacent points. This would help distinguish meaningful signal peaks from noise.

9.3.6 Predictive and Interpolative Techniques

Significant point extraction is a particular data compression method, related to the generalized techniques based on *predictors* and *interpolators*. These are algorithms that operate on waveforms or other data streams and produce compression by reducing the amount of redundancy present in that data. As long as the data set is not random, there is some correlation between adjacent data values that can be exploited.

Predictive encoding techniques use the information contained in previous data samples to extrapolate (or predict) the value of the next data sample. This approach is used extensively in data communications systems for compressing data streams "on the fly," just prior to transmission (often using dedicated hardware). This extrapolation is done by fitting a function (or polynomial) to the existing data. Usually, only a zero-order (constant) or first-order (linear) function is used, since high-order functions tend to be very sensitive to noise and can become unstable.

The simplest extrapolation method is the *zero-order* predictor with a fixed aperture. In Figure 9-7a, a sample waveform is shown with its discrete points. Starting with the first data point, a vertical aperture (or window) of fixed amplitude $2d$ is drawn around the first point. Additional $2d$ windows are extended over the full amplitude of the waveform. The first point is always saved, and saved points are denoted by the X character. If the next point's amplitude fits within the same $2d$ window, it is discarded, otherwise it is saved. After determining a new point to save, subsequent points that fit within the new $2d$ window are discarded. Of course, the x coordinate (usually time) of the saved points must also be kept.

Figure 9-7b shows the reconstructed waveform, using only the saved points from Figure 9-7a. Notice how using a zero-order predictor tends to "flatten out" small amplitude changes. Obviously, there is a moderate amount of data distortion using this technique.

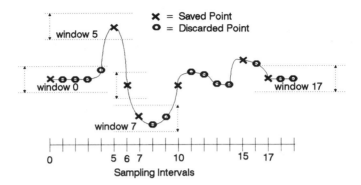

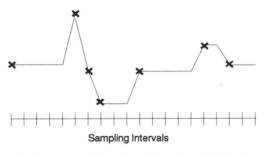

Figure 9-7 Zero-order predictor (ZOP) used for waveform data compression.

Data compression can be improved using a zero-order predictor with a *floating aperture*. Instead of the window locations being fixed by the value of the first data point, each new $2d$ aperture is centered around the last point saved. In this case, if a new point is close in amplitude to the last saved point it will always be discarded. With a fixed aperture, if this new point happened to be just over the next aperture boundary, it would be unnecessarily saved.

An approach more flexible than the zero-order predictor is the *first-order predictor* (FOP) or the *linear predictor*. This is a very popular method used for many applications, such as compressing digitized human voice data. For this use, some data distortion is acceptable, since the final receiver (a human being) can still understand moderately garbled data.

Using a linear predictor is very similar to implementing a zero-order predictor, except now new data points are predicted by extrapolation from a line connecting the previous two points. Figure 9-8a shows the same sample waveform as in Figure 9-7a. The points saved by the algorithm are again marked with the character X. The first two points are always saved, to generate the first line. The following three points fit on the line, within the error window of $2d$; they can be discarded, since a reconstruction algorithm can extrapolate them from that line. The next point does not fit within the line and must be saved. A new line is drawn between this newly saved point and the previous, extrapolated point. The

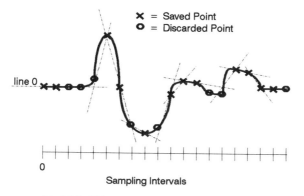

(a) Original Waveform and Sampled Points Using Linear Predictors

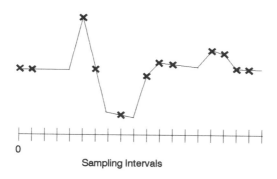

(b) Reconstructed Waveform from Saved points

Figure 9-8 First-order (linear) predictors used for waveform data compression.

next point does not fit on this line and is saved, generating another line that following points do fit. This process continues, discarding points that fit (within $\pm d$) existing extrapolation lines and saving those that do not, while drawing new lines.

When the resulting saved points reconstruct the waveform in Figure 9-8b, we see that more of the fine details and curvature of the original waveform are maintained by the linear predictor, compared to the zero-order predictor. The compression ratios from both techniques are also comparable.

When data does not have to be compressed in real time, if it has been previously acquired and stored, interpolator techniques can be used. These are very similar to the predictor methods, except that now interpolation is used instead of extrapolation.

For example, using a *linear interpolator* is very similar to using a linear predictor. Using the waveform in Figure 9-9 as an example, the first point is always saved. The second point is skipped, and an imaginary line is drawn from the first to the third point. If the second point falls on this line within a $2d$ window, it is discarded. A new line is tested between the first and fourth points. If both the second and third points fall on this line (within the tolerance window of $2d$), they are both discarded. This process continues until a line is drawn that does not fit all the intermediate points. The last point that ended an acceptable test line (the fifth point, ending line 1 in this example) is saved. For data reconstruction, the intermediate, discarded points are interpolated between the two saved end points. Now, the process starts again with the end point of the last line serving as the start point for a new line. When this process is complete, at the last point in the waveform, the saved points represent the end points of interpolation lines used for reconstructing the data.

Sometimes, no intermediate points can be discarded and adjacent points are saved, especially at the peak of a curve. Since this approach requires the entire waveform present before processing can occur, it is not suitable for real-time compression. It is very useful for post-acquisition or post-processing applications. As with a linear predictor, a linear interpolator does produce data distortion. This can be balanced against the compression ratio by adjusting the window size. A larger window will produce higher distortion along with a higher compression ratio. Typically, an interpolator will produce a higher compression ratio than an equivalent predictor, with slightly less distortion.

Since all predictors and interpolators produce an output array of (x,y) points, they are often combined with other techniques, such as delta modulation and Huffman encoding, to reduce the total number of bits required to store the compressed waveform. The true measure of the

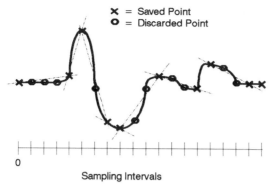

(a) Original Waveform and Sampled Points Using Linear Interpolators

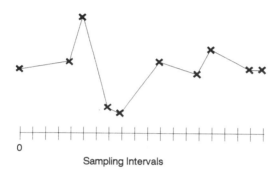

(b) Reconstructed Waveform from Saved points

Figure 9-9 First-order (linear) interpolators used for waveform data compression.

compression ratio for the overall process is its bit compression ratio (as opposed to the point compression ratio, produced by the predictor or interpolator alone):

$$\text{Bit compression ratio} = b_0/b_c$$

where

b_0 = number of bits in original waveform
b_c = number of bits in compressed data

Quite often, the optimum compression technique for a particular class of data must be determined strictly by trial and error. The data compression information in this chapter is hardly exhaustive. Certain nonlinear curve fitting techniques, such as splines, are commonly used. Fields that use extremely large data sets, such as imaging, have numerous, dedicated compression techniques that produce very large compression ratios. Many commercial data-compression products are available for use on PCs. Some are hardware-based, for increasing hard disk storage without utilizing CPU overhead. Other products are strictly software-based, usually for producing hard-disk file backups. Since the nature of the data stored on a PC's files can vary tremendously, intelligent systems can determine the compression algorithm to use based on the data itself.

This concludes our look at PC file storage and data compression. In the next chapter we will examine some common processing and analysis techniques applied to acquired data, along with considerations of numerical representation and precision.

CHAPTER 10

Data Processing and Analysis

The power and flexibility in using a personal computer as a data acquisition platform is shown most clearly by how data can be manipulated once it is acquired. In this chapter we will explore some of the data analysis and processing techniques commonly used with data acquisition systems.

Since most data collected by data acquisition systems are numeric, it is important to know how numbers are represented and manipulated on a computer. We will start by looking at numerical representation and storage on a personal computer.

10.1 Numerical Representation

As we previously touched on while discussing ADCs and DACs, there are many possible ways to represent conventional decimal numbers in a binary format. The simplest of these are integer representations. For nonintegral numbers, various fractional formats can be used, though for maximum flexibility and dynamic range, floating-point representations are preferable.

10.1.1 Integer Formats

The fastest and most efficient way to manipulate data on a personal computer is to store it in an integer format. An integer can be either signed (representing both positive and negative numbers) or unsigned (positive numbers only). The maximum dynamic range of the values that can be represented is determined by the number of bits used. Therefore, n bits can represent 2^n numbers with a dynamic range (in dB) of $20 \log_{10}(2^n)$. If $n = 8$, then 256 different integers can be represented: positive integers in the range 0 to 255, or signed integers in the range -128 to $+127$. This corresponds to a dynamic range of 48 dB. If $n = 16$ bits, 65,536 values can be represented, for a dynamic range of 96 dB.

The standard integer formats commonly used on a PC are byte (8 bits), word (16 bits), long word (32 bits), and double word (64 bits), as shown in Table 10-1. On an Intel 80×86 family PC (PC/XT/AT), data is addressed on a byte-by-byte basis. The starting memory address for a word (or long word) is the first of the two (or four) bytes comprising that word. The first (addressed) memory location contains the least significant byte (LSB), while the last location contains the most significant byte (MSB), as illustrated in Figure 10-1.

This byte ordering is processor-dependent. On a computer based on a Motorola 68000 series CPU, such as an Apple Macintosh, a different storage arrangement is used. All words must start at an even address with the MSB at the starting (even) address and the LSB at the higher (odd) address. For a long word, the high-order 16 bits are stored at the starting (lower) address and the low-order 16-bits at the higher address (start $+2$).

Most of the time, the method used by a CPU to store and access data in memory is transparent to the user and even the programmer. It only becomes an issue when one data storage element, such as a word, is also accessed as a different element, such as a byte. Due to the strong likelihood of error in doing this, it is not a recommended approach.

The nature of data storage depends only on how many bytes are needed to represent a particular data storage element. An unsigned integer is usually represented as a natural binary number, such as $25 = 11001$. If an element is a signed integer, there are several ways to encode or

TABLE 10-1
Integer Formats

INTEGER TYPE	# OF BITS	SIGNED VALUES	UNSIGNED VALUES
Byte	8	-128 to +127	0 to 255
Word	16	-32768 to +32767	0 to 65535
Long Word	32	-2.14×10^9 to $+2.14 \times 10^9$	0 to 4.29×10^9
Double Word	64	-9.22×10^{18} to $+9.22 \times 10^{18}$	0 to 1.84×10^{19}

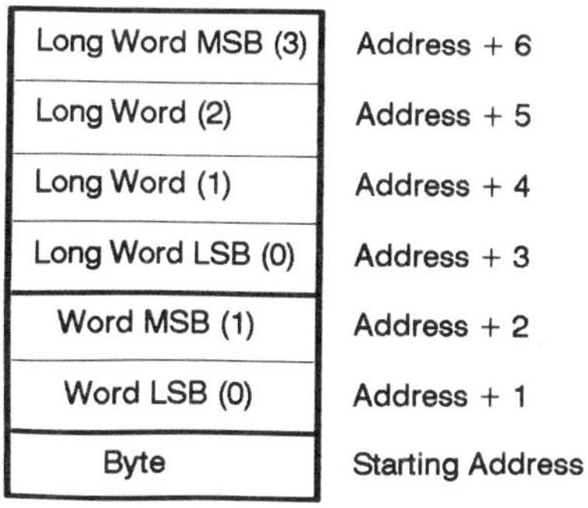

```
| Long Word MSB (3) | Address + 6     |
| Long Word (2)     | Address + 5     |
| Long Word (1)     | Address + 4     |
| Long Word LSB (0) | Address + 3     |
| Word MSB (1)      | Address + 2     |
| Word LSB (0)      | Address + 1     |
| Byte              | Starting Address|
```

Figure 10-1 Multibyte integer storage in 80×86 PC memory.

represent it. The most popular approach is to use twos-complement representation, as shown in Table 10-2. Using twos-complement representation, the most significant bit is used as a sign bit. If it is 0, the number is a positive integer, with the same value as its unsigned binary counterpart. If the sign bit is 1, the number is negative.

The twos-complement value is calculated by first writing the binary value of the corresponding positive number, then inverting all the bits, and finally adding 1 to the result. To get the 4-bit twos-complement representation of −4, we start with the unsigned binary value for +4 = 0100. When we invert all the bits, we get 1011. Adding 0001 to this number produces the final value of 1100 = −4. Using twos-complement representation for negative integers is widely accepted because if you add corresponding positive and negative numbers together using this system, you will get a result of zero, truncated to the original number of bits. So, adding −4 to +4, we get:

$$\begin{array}{r} 1100 \\ +0100 \\ \hline =0000 \end{array}$$

The use of twos-complement representation gives us the n-bit signed integer range of $-2^{(n-1)}$ to $+2^{(n-1)} - 1$ (i.e., for $n = 4$ this range is −8 to +7).

Other encoding techniques are used to represent decimal integers in a binary format besides natural binary and twos complement. One of the more common alternatives is binary coded decimal (BCD). This code uses four bits to represent a decimal digit, in the range 0 to 9. It uses natural unsigned binary representation (0000 to 1001). The six codes above 9

TABLE 10-2
Four-bit Signed Integers

DECIMAL VALUE	TWOS COMPLEMENT BINARY CODE
+7	0111
+6	0110
+5	0101
+4	0100
+3	0011
+2	0010
+1	0001
0	0000
-1	1111
-2	1110
-3	1101
-4	1100
-5	1011
-6	1010
-7	1001
-8	1000

(1010 to 1111 or Ah to Fh) are unused. To represent a decimal value, a separate BCD code is used for each decimal digit. For example, to represent the value 437:

$$437 = \underset{(4)}{0100} \ \underset{(3)}{0011} \ \underset{(7)}{0111}$$

If only one BCD digit is stored in a byte (upper four bits are set to 0), it is called unpacked BCD storage. If two BCD digits are stored in a byte it is called packed BCD storage. Even using packed storage, BCD numbers require more storage than natural binary or twos-complement values. As an illustration of unsigned integers, four BCD digits (16 bits) can represent the values 0 to 9999, and a natural binary word (16 bits) can represent values 0 to 65,535. Alternately, we only need 16 bits to represent 50,000 with an unsigned natural binary word, while we need 20 bits (five digits) to do the same with BCD. BCD is popular with systems processing large amounts of important numerical data, such as those used by financial institutions.

An even less efficient means of numerical representation is using an ASCII character to represent each decimal digit. In this case, seven (or eight) bits are needed to represent 0 to 9 (as well as sign and decimal point, for nonintegers). This is about twice as inefficient as BCD representation. ASCII numerical representation is usually used strictly to store data in a format that is easy to read and print. It is usually converted into a format more convenient to use before numerical processing can proceed.

10.1.2 Noninteger Formats

Quite often, when a computer is used to process acquired data, integer precision is not adequate, either due to round-off errors, dynamic range limitations, or poor modeling of the measured phenomena. Several numerical formats are used to overcome this problem.

The simplest way to depict fractional values is with fixed-point representation, which is basically an extension of binary integer representation. For integer representation using n bits, the binary number $b_n b_{n-1} \ldots b_1 b_0$ is evaluated by adding the weighted value of each nonzero bit as follows:

$$b_n * 2^n + b_{n-1} * 2^{n-1} + \cdots + b_1 * 2^1 + b_0 * 2^0$$

where, b_i is the ith bit (0 or 1). For binary fixed-point representation, both positive and negative exponents are used and a binary point appears after the 2^0 digit. For example, if we had an 8-bit number with a 3-bit fraction, it would be written as

$$b_4 b_3 b_2 b_1 b_0 \cdot b_{-1} b_{-2} b_{-3}$$

The weights for bits b_{-1}, b_{-2}, and b_{-3} are 2^{-1}, 2^{-2}, and 2^{-3}, respectively. The resolution of this representation is 0.125 (2^{-3}), and its range of values for unsigned numbers is 0 to 31.875, which is still the same number of values as an 8-bit unsigned integer (31.875/0.125 = 255).

When more bits are added to unsigned integers, the resolution stays the same (1) while the range of values increases. When the number of bits after the binary point in a fixed-point, fractional representation increases, the resolution increases, while the range of values stays the same. This tradeoff between range of values and resolution is inherent in these representations.

If we needed to increase both the dynamic range and resolution of our numerical representation, we could keep increasing the number of bits per number. The problem here is that most CPUs can perform math on only a fixed number of bits at a time. For 16-bit processors (as used in many personal computers), if more than two bytes represent a number, additional instructions must be performed when executing a math function, splitting the function into multiple 16-bit operations. If we are using 32-bit integers and need to add them, we have to first add the lower 16-bit words, then add the upper 16-bit words with any carry from the previous addition. The software overhead and processing time increases quickly as we increase the size of numerical elements.

The standard solution to this dilemma is to use a floating-point format, consisting of a fractional part (the *mantissa*) and an *exponent*. The number of bits used to represent the exponent determines the floating-point number's range of values, and the number of bits used for the mantissa determines its resolution. The mantissa is a signed, binary fraction that is multiplied by 2^{exp} to produce the represented value. The exponent is a signed integer.

Certain standard formats are used to represent floating-point numbers. Among the most popular, the IEEE 754 Floating-Point Standard is also commonly used with personal computers. This standard defines two formats: single precision, using 32 bits, and double precision, using 64 bits, as shown in Figure 10-2.

In both formats the sign bit (most significant bit) is for the mantissa, which is in a normalized form (with a value between 1.0 and 2.0). In fractional binary, this would be 1.000 ... 0 through 1.111 ... 1 (using a

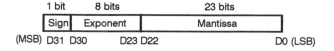

(a) Single Precison (32 bits)

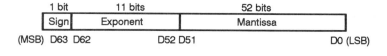

(b) Double Precison (64 bits)

Figure 10-2 IEEE floating-point formats.

10.1 Numerical Representation 177

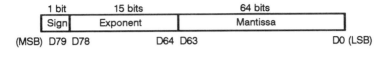

Figure 10-3 Intel 8087 80-bit temporary floating-point format.

fixed binary point). Since the most significant mantissa bit (before the binary point) is always 1, it is implied and not stored with the number. So, a single precision mantissa of 1.01101111000010101010011 would be stored as 01101111000010101010011.

The exponent is stored in a *biased* form, with a fixed value, or bias, added to it. For single-precision numbers, this bias is +127, and for double-precision numbers it is +1023. This biased exponent is useful for determining which of two exponents is larger by comparing them bit by bit, starting with the leftmost bit. For example, consider two single-precision numbers with exponents of +15 and −5, represented as signed integers:

$$+15 = 00001111 \quad -5 = 11111011$$

and represented as biased integers (+127):

$$+15 + 127 = 10001110 \quad -5 + 127 = 01111010$$

So, just looking at the leftmost bit indicates that +15 is the larger exponent.

The valid exponent range for single precision is −126 to +127, and for double precision it is −1022 to +1023. When represented as a biased exponent, a value consisting of either all 0's or all 1's indicates an invalid number. This way numerical overflow/underflow errors can be indicated.

A special, non-IEEE format is used on Intel 80×86 family PCs with 80×87 math coprocessors, the *temporary format*, shown in Figure 10-3. This is an 80-bit format, incorporating a 64-bit mantissa with a 15-bit exponent. It is very useful for highly repetitive mathematical operations where round-off errors can reduce precision, as well as calculations involving very large or very small numbers.

The temporary format uses an exponent bias of +16383. It differs in spirit from the single and double-precision formats by explicitly keeping the leftmost 1 in the normalized mantissa value. Since the math operations using this 80-bit temporary format are performed in hardware, the large number size does not cause severe processing speed penalties. The Intel math coprocessors are discussed in further detail in Chapter 12.

TABLE 10-3
Range and Precision of Various Numerical Formats

NUMBER TYPE	FORMAT	TOTAL # OF BITS	EXPONENT/ MANTISSA BITS	DECIMAL DIGITS OF PRECISION	DECIMAL RANGE
INTEGER	BYTE	8	—	>2	$>+/-10^2$
	WORD	16	—	>4	$>+/-10^4$
	LONG WORD	32	—	>9	$>+/-10^9$
	DOUBLE WORD	64	—	>18	$>+/-10^{18}$
FLOATING POINT	SINGLE PRECISION	32	8 / 23	> 7	$>+/-10^{38}$
	DOUBLE PRECISION	64	11 / 52	> 15	$>+/-10^{307}$
	TEMPORARY	80	15 / 64	> 19	$>+/-10^{4932}$

Table 10-3 lists decimal precision (number of significant digits) and range for some of the integer and floating-point numerical formats we have discussed here. Note that for an equivalent number of bits (such as 32 or 64), floating-point formats have slightly lower precision along with much higher dynamic range than the corresponding integer formats. This is simply due to diverting some of the bits used for precision in an integer to the exponent of a floating-point value, increasing its range.

10.2 Data Analysis Techniques

A wide variety of processing techniques are commonly applied to the data produced by data acquisition systems. These can range from simply plotting the data on a graph to applying sophisticated digital signal processing (DSP) algorithms. A large number of commercial software packages, such as those discussed in the next chapter, have many of these capabilities built in. This enables the user to concentrate on the data analysis without getting bogged down in the details of programming a personal computer. We will begin our survey of data processing by looking at statistical analyses.

10.2.1 Statistical Analysis Techniques

The most common analysis applied to acquired data is some statistical calculation. Statistical parameters describe the distribution of values within a data set. They indicate where data values are most likely to be found as well as the probable variability among them.

The most important statistical measurement for a data set is the mean, which is simply the average of a set of values. If we have a set Y of n values, $y_1, y_2, \ldots, y_n$, the mean of Y is just:

$$y_m = (y_1 + y_2 + \cdots + y_n)/n$$

The values of the data set must have some relationship to each other for the mean to have significance. For example, the set may consist of n measurements of the same quantity, repeated over time.

The conventional mean is used to analyze an existing data set. A special variation on the mean is the *running average*, sometimes referred to as the circular average or sliding average. The running average is useful for real-time control applications, when the current average value is needed. For instance, an intelligent heater controller needs to know the current temperature of a system to apply the appropriate amount of heater power. If the temperature varies significantly from reading to reading, an average of the last n readings would be useful to smooth out this temperature noise. The running average is just the mean of the last n values. If the current reading is T_i and the running average is n points wide, its value at point i is

$$T_{mi} = (T_i + T_{i-1} + \cdots + T_{i-n+1})/n$$

At the next point, $i + 1$, the running average is

$$T_{mi+1} = (T_{i+1} + T_i + \cdots + T_{i-n+2})/n$$

The running average is updated with each new value acquired. It acts as a low-pass filter on the incoming data. Only relatively slow artifacts with large amplitude changes will be reflected in the running average. When this technique is applied to an existing, acquired waveform, the n-point averaging window is usually symmetric around the selected point.

Another statistical measurement is the *median*. It is selected so that half of the data set values are higher than the median and the other half are lower. The median is often close in value to the mean, but it does not have to be.

An important measure of variation within a set of data is the *standard deviation*. If we have a data set $(y_1, y_2, \ldots, y_n)$ of n values with a mean of y_m, the standard deviation σ is

$$\sigma = [[(y_1 - y_m)^2 + (y_2 - y_m)^2 + \cdots + (y_n - y_m)^2]/n]^{1/2}$$

This is a measure of the differences between the data set values and the average value. The smaller the standard deviation, the "tighter" the distribution of data values. In the case where all values in a data set are identical, the standard deviation would be zero.

When a data set fits a normal Gaussian distribution (a "bell" curve), approximately 68% of the values will be found within one standard deviation of the mean value. As an illustration, assume a manufacturer is interested in analyzing the length of a production part. Length measurements are taken on a sample of parts that fit a Gaussian distribution, having as its peak the mean value. Here, the standard deviation is a measure of the length variations from part to part. From these measurements, the manufacturer can predict the percentage of a production run that will fall within an acceptable tolerance. If this percentage is too small, it indicates a need to control the production process better.

10.2.2 Curve Fitting

The mean and standard deviation are mostly used on sets of values that should be describing the same or similar measurements. When the acquired data is a waveform, described as two-dimensional (x,y) points, a common requirement is comparing it to a theoretical model, or finding a model that fits the data. This data can be a one-dimensional array where time is the independent (x) variable. The theoretical model describes a waveform that should be similar to our acquired data. Finding a mathematical model that fits the measured data is referred to as curve fitting.

Very often, a polynomial is used to describe a theoretical curve. The general form of a polynomial of order n is

$$F(x) = a_0 + a_1 x + a_2 x^2 + \cdots + a_n x^n$$

The coefficients are the constants a_i, which are adjusted during the curve fitting process.

To determine the coefficient values for the best curve fit, the sum of the error terms for each of j data point in the curve is calculated as

$$[F(x_1) - y_1]^2 + [F(x_2) - y_2]^2 + \cdots + [F(x_j) - y_j]^2$$

where $y_1, y_2, \ldots, y_j$ are the measured values. When this function is minimized, the coefficients for $F(x)$ describe the best curve fit to the data set. This is referred to as the least squares fit. Using least squares to test how well a function fits a data set is not limited to polynomials. Exponential and trigonometric functions are also commonly employed for curve fitting and still use a least squares fit measurement. The iterative type of calculations used to find the least squares fit is well suited to digital computer calculations.

The simplest curve fitting is a first-order ($n = 1$) or linear fit, sometimes referred to as a linear regression. Analytically, the coefficients a_0 and a_1 are determined. Graphically, a straight line is drawn through the data points. The resulting line is described by the standard formula:

$$y = mx + b$$

where m is the slope and b is the y-intercept. This means that the general coefficients $a_0 = b$ and $a_1 = m$.

Given a set of (x,y) data points, the coefficients for a linear regression can be determined analytically. The coefficients a_0 and a_1 are calculated from summations over all n points in the data set:

$$a_0 = \frac{[(\Sigma\, y) * (\Sigma\, x^2) - (\Sigma\, x) * (\Sigma\, xy)]}{[n * (\Sigma\, x^2) - (\Sigma\, x)^2]}$$

$$a_1 = \frac{[n * (\Sigma\, xy) - (\Sigma\, x) * (\Sigma\, y)]}{[n * (\Sigma\, x^2) - (\Sigma\, x)^2]}$$

For higher-order polynomial fits, analytic approaches are impractical. An iterative process of successive approximations is typically used.

Figure 10-4 is a simple example of a linear curve fit. There are four (x,y) values: (1,2), (3,4), (5,5), and (8,6). Calculating the coefficients from the above equations, we find the least squares fit line to these points is $y = 0.55x + 1.9$.

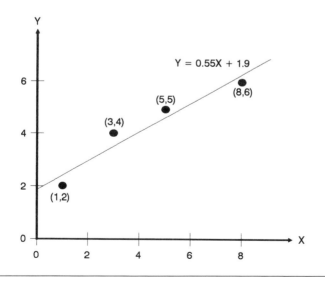

Figure 10-4 Example of linear curve-fitting.

Notice the similarity between linear curve fitting and linear predictors/interpolators, discussed in Chapter 9. In both cases, a straight line is found that best fits the data. Furthermore, minimizing the distortion produced by data compression is often a least squares process.

Curve fitting is a broad, complex field. This brief discussion should give you a feel for implementing curve fitting on a personal computer-based data acquisition system. An advantage of using these systems (with appropriate software) is the ability to see the data graphically, along with getting the numerical processing power of a personal computer. When it comes to processing waveforms, seeing the data displayed as a graph is invaluable.

10.2.3 Waveform Processing

A large portion of the information gathered by data acquisition systems is in the form of waveforms (commonly, a function varying with time). These waveforms are easily displayed graphically, using many of the software packages described in the next chapter. Very often, this acquired data is operated on as a single entity: a vector (one dimension) or array (two or more dimensions). Many of these operations are simple mathematical functions, such as subtraction or multiplication with a scalar or another array.

Consider the example in Figure 10-5a, a waveform representing an ultrasonic pulse, which should have a net DC component of zero. Due to DC offsets in the analog receiver system, the acquired signal may not meet this criterion. To determine the net DC offset, we take the mean value of all the waveform points. If this mean is not zero, we subtract it (a scalar) from the waveform (a vector). The result, shown in Figure 10-5b, now has a zero DC offset.

Waveforms can also be used to operate on each other. For example, special windowing functions are commonly used in DSP algorithms. Waveforms under analysis are multiplied by these windowing functions, which are also waveforms. In many cases a reference or baseline waveform is acquired. Subsequent data is then divided by this reference data, for normalization.

Other common operations are integration and differentiation. If we wish to determine portions of a waveform with high slopes, we would differentiate it. The peaks of a differentiated function occur at slope maximums. When a particular function is difficult to differentiate or integrate analytically, this numerical approach is very useful. For numerical differentiation, the slope dy/dx is calculated for every pair of adjacent points.

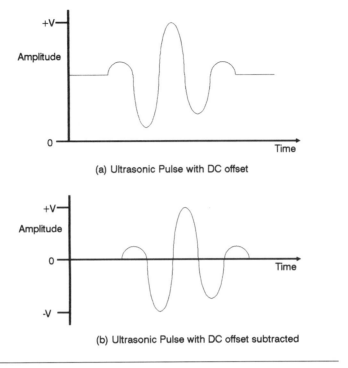

Figure 10-5 Example of subtracting a DC offset from a waveform.

In a similar fashion, the area under the curve at each point is calculated for numerical integration.

For example, suppose a waveform represented the measured displacement of an object versus time. Differentiating this waveform would produce a new waveform representing the object's velocity versus time. Differentiating a second time would produce an acceleration-versus-time waveform. Conversely, if the acquired waveform represented acceleration data, as from an accelerometer, integrating it once would produce a velocity curve and integrating it a second time would produce a displacement curve. The only problem here is that any fixed offsets in either displacement or velocity would not appear in the integrated data, as they were lost by the original acceleration measurements.

Again, this brief discussion is only scratching the surface on the topic of waveform processing. Many mathematical operations are performed on data representing vectors and arrays, such as dot products and cross products. The huge variety of waveform processing techniques find an immense range of applications. We will look at a few specialized techniques now, starting with Fourier transforms.

10.2.4 Fourier Transforms

Undoubtedly, Fourier transforms are among the most popular signal-processing techniques in use today. Analytically, the Fourier series for a single-valued periodic function is a representation of that function using a series of sinusoidal waveforms of appropriate amplitude and phase. The sine waves used in the series are at multiple frequencies (harmonics) of the lowest frequency (the fundamental). The Fourier series for a periodic function $f(t)$, with a period T would be

$$f(t) = a_0 + a_1 \sin(\omega t + \phi_1) + a_2 \sin(2\omega t + \phi_2)$$
$$+ \cdots + a_n \sin(n\omega t + \phi_n)$$

where

$\omega = 2\pi/T$, the fundamental frequency;

$a_1, \ldots, a_n$ are the amplitude values for each frequency component (a_0 is the DC component);

$\phi_1, \ldots, \phi_n$ are the phase values for each frequency component.

To represent a single-frequency sinusoidal wave, only the DC and fundamental frequency terms are needed. Most functions require many terms to provide a good approximation of their real value. For example, Figure 10-6 shows a square wave, which has a Fourier series consisting of decreasing odd harmonics:

$$f(t) = \frac{4a_0}{\pi} \left[\sin(\omega t) + \frac{1}{3} * \sin(3\omega t) + \cdots + \frac{1}{n} * \sin(n\omega t) \right]$$

Using only the first term (fundamental frequency) we get a crude approximation of the real waveform. After we use the first three terms (up to the fifth harmonic) we have a much closer approximation of the square wave.

By fitting trigonometric functions to an arbitrary waveform, we can get the frequency content of that waveform. In essence, the Fourier transform is used to convert from a conventional data (time) domain waveform to a spectral (frequency) domain waveform. Since this transformation is bilateral, an inverse Fourier transform converts data back from the frequency domain into the time domain. Data domain waveforms include functions of both time and space. The Fourier transform of a distance-based waveform contains spatial frequency information.

Analytically, the Fourier transform is defined for operation on continuous, periodic functions. Given a function of a real variable (the func-

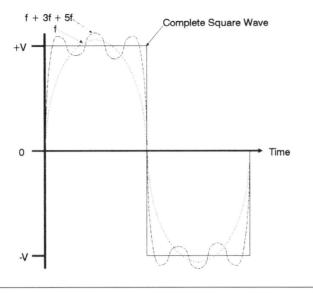

Figure 10-6 Fourier series for a square wave.

tion itself can be complex) $f(x)$, its continuous Fourier transform (CFT) $F(y)$ is defined as

$$F(y) = \int_{-\infty}^{\infty} [f(x) * e^{-j2\pi xy} \, dx]$$

This integral must exist for every real value of x. The complex exponential used in the integral has an equivalent trigonometric form, using Euler's formula:

$$e^{jx} = \cos(x) + j \sin(x)$$

where $j = \sqrt{-1}$, the imaginary number operator.

An alternate form for the CFT would be

$$F(y) = \int_{-\infty}^{\infty} f(x)[\cos(2\pi xy) - j \sin(2\pi xy)] \, dx$$

For data acquisition applications, a special Fourier transform is used to operate on discrete, finite functions. This is called the discrete Fourier transform (DFT) and is used to operate on discrete (digitized) data. The DFT is the workhorse of DSP techniques. If we have a waveform $f(k)$ consisting of n points, the DFT produces a complex waveform of n points, $F(m)$. Both k and m vary from 0 to $n - 1$. The data points of $f(k)$ are evenly spaced in the time domain by dt; they range from 0 to $(n - 1)\, dt$. The transformed data points of $F(m)$ are evenly spaced in the frequency domain by $1/dt$ and range from 0 to $(n - 1)/dt$. The DFT is calculated from

$$F(m) = \sum_{k=0}^{n-1} [f(k) * e^{[(-j2\pi/n)mk]}]$$

The frequency-determining component is $2\pi m/n$, which is a normalized value. The DFT assumes the time domain waveform is a periodic function, with a period of n points. The normalized frequency at the first DFT point is 0 and at the last point is $2\pi(n - 1)/n$ radians. This maximum frequency is $(n - 1)/dt$, so the time domain sampling is normalized to $dt = n/2\pi$.

Note that the first term of the DFT, $F(0) = \Sigma\, f(k)$, at zero frequency ($m = 0$). This is simply the area under the curve or the result of integrating $f(k)$. Also note that for each term in $F(m)$, n complex multiplications must be done of $f(k)$ times the complex exponential term [where $f(k)$ can be either real or complex]. It is a fair assumption that the amount of time required to calculate a DFT using a digital computer is proportional to the number of complex multiplications (each involving four separate real multiplications and additions). Since n complex multiplications are performed for each of n points, the number of complex multiplications required to perform a DFT is proportional to n^2. As the number of input points n increases, the time required to calculate the transform goes up by the square. When real-time frequency analysis is required on a large amount of data, such as with spectrum analysis, the required computation time can be much too long. In this case, the output frequency data (DFT) falls behind the input time domain data.

If we have frequency-domain data and want to convert it back to the time domain, we can use the inverse DFT:

$$f(k) = \frac{1}{n} * \sum_{m=0}^{n-1} [F(m) * e^{[(j2\pi/n)mk]}]$$

For the inverse transform, the frequency data $F(m)$ is multiplied by a complex exponential and summed over all its points to calculate each $f(k)$ point. Notice the scale factor of $1/n$ here. As with the forward DFT, the time required to compute the inverse DFT is proportional to the square of the number of points.

The answer to the problem of DFT computations taking too long to calculate is the fast Fourier transform (FFT), which is a special implementation of the DFT. By exploiting the symmetry inherent in the DFT and breaking up the calculations into several smaller transforms, the FFT computation time can be greatly reduced. Most FFT algorithms only operate on a set of points that is an exact power of 2 ($n = 2^x$). However, the number of complex multiplications required by an FFT is only $n \times \log_2(n)$. So an FFT is $n/\log_2(n)$ faster than an equivalent DFT. For a waveform of 1024 points, this is a speed-up of over 100 times (1024/10).

For the rest of this discussion, we will assume that the Fourier transforms used on a personal computer will always be FFTs. The commercial software packages listed in Chapter 11 (and Appendix B) that contain Fourier transform functions all employ an FFT algorithm.

Some of the symmetry inherent in the FFT of a waveform is shown by plotting it. All FFTs are complex waveforms with a real and imaginary component for each frequency value (point). If the original time-domain function is real, the real component of its FFT has even symmetry (symmetrical about point $n/2$) and the imaginary component has odd symmetry (antisymmetric about $n/2$). If the original function is imaginary, the real component of the FFT has odd symmetry and the imaginary component has even symmetry. If the original function is purely real or purely imaginary, the magnitude of its FFT will have even symmetry.

Very often, when looking at the FFT of a waveform for frequency analysis, only the magnitude $|F(m)|$ is of interest. Since the FFT points are complex:

$$|F(m)| = [(F(m)_{\text{real}})^2 + (F(m)_{\text{imag}})^2]^{1/2}$$

If the signal of interest in the time domain is an ideal impulse, infinitely sharp (all but one point is zero amplitude), the magnitude of its FFT is a constant. That is, an impulse contains a spectrum of equal amplitude at all frequencies. This makes an impulse very useful as a broad-band excitation signal.

As an example, Figure 10-7a shows a simple rectangular pulse of unit amplitude (1.0), eight points wide in a 64-point waveform. Figure 10-7b displays the magnitude of the FFT of this simple waveform.

Notice the even symmetry of the FFT magnitude. This is because the original function was purely real. For an FFT of n points, the magnitude is symmetrical about point $n/2$. The actual frequency data is valid only up to point $n/2$, which is half the entire frequency range. Since the maximum frequency is equal to the original data acquisition sampling rate ($f_s = 1/dt$, where dt is the time between consecutive samples) the FFT data is valid only up to $f_s/2$, the Nyquist frequency. Above that point it is just the mirror image.

Another interesting feature is the periodicity in the magnitude of the FFT, displayed in Figure 10-7b. With a rectangular pulse y points wide in

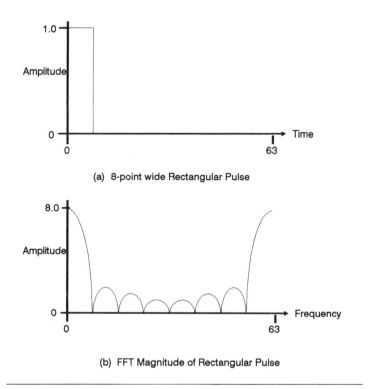

Figure 10-7 Example of fast Fourier transform (FFT): 64-point FFT of 8-point-wide rectangular pulse.

the time domain, the period in the frequency domain is n/y, which is every eight points in this case. If the rectangular pulse were wider, the number of peaks in the FFT magnitude would increase as the period decreased. Also, note that the value of the zero-frequency point $|F(0)| = 8$. This is equal to the value obtained by integrating the original pulse waveform (eight points wide with an amplitude of 1), which is its DC component.

Figure 10-8a displays a 64-point exponential waveform decay, e^x, from e^1 at point 0 to $e^{(1/64)}$ at point 63. The magnitude of its FFT is shown in Figure 10-8b. Again, the value we get for $|F(0)|$ is equivalent to the result of integrating under the waveform, which has a large DC offset (note that the exponential waveform does not approach a zero value in the sampled time interval).

The following is a simple FFT program written in BASIC. It will run under IBM BASIC or GW BASIC. Since BASIC is an interpreted language (see Chapter 13 for more details), it executes slowly. The actual

FFT or (IFFT) computation is done by the subroutine starting at line 400. The test program, starting at line 10, allows the user to enter a 16-point data array as input to the FFT subroutine. This program is only useful for relatively small data arrays, such as 64 points or fewer. For larger arrays, depending on the personal computer used, the FFT computation time could take several minutes.

```
10    REM - FFT PROGRAM, TESTS FFT SUBROUTINE WITH
20    REM - ARRAY OF 16 POINTS, PROVIDED BY USER.
30    PI = 3.14159
40    N = 16            'NUMBER OF POINTS IN WAVEFORM
50    DIM R(N)          'REAL DATA ARRAY, INPUT & OUTPUT
60    DIM I(N)          'IMAGINARY DATA ARRAY, INPUT & OUTPUT
70    PRINT "      FFT TEST PROGRAM": PRINT
80    PRINT "NUMBER OF POINTS = "; N: PRINT
90    INPUT "REAL DATA INPUT, ONLY - Y OR N?   ",A$
100   CLS               'CLEAR SCREEN
110   INPUT "INPUT DELTA T (1):   ",DELTA
120   PRINT "INPUT SIGNAL DATA POINTS" : PRINT
130   FOR J = 1 TO 16
```

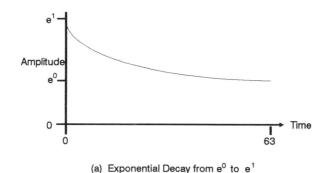

(a) Exponential Decay from e^0 to e^1

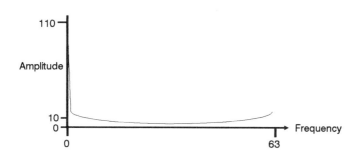

(b) FFT Magnitude of Exponential Waveform

Figure 10-8 Example of 64-point FFT of exponential decay waveform.

```
140   PRINT "POINT "; J; ": ";
150   INPUT "XR = ",R(J)
160   IF A$ = "Y" THEN I(J) = 0! : GOTO 190
170   PRINT "POINT "; J; ": ";
180   INPUT "XI = ",I(J)
190   PRINT
200   NEXT J              'END OF DATA INPUT LOOP
210   PRINT
220   CLS                 'CLEAR SCREEN
230   PRINT "CALCULATING FFT ...................."
240   GOSUB 400           'CALL FFT SUBROUTINE
250   PRINT: PRINT "POINT","FFT REAL","FFT IMAG"
260   FOR J = 1 TO N
270   PRINT J,R(J),I(J)
280   NEXT J
290   PRINT
300   INPUT "DISPLAY FFT MAGNITUDE & PHASE - Y OR N?  ",A$
310   IF A$ < > "Y" THEN STOP
320   PRINT: PRINT "POINT", "FFT AMP", "FFT PHS"
330   FOR J = 1 TO 16
340   AMP = (R(J)^2 + I(J)^2)^.5
350   PHS = PI/2
360   IF R(J) < > 0 THEN PHS = ATN(I(J)/R(J))
370   PRINT J,AMP,PHS
380   NEXT J
390   STOP
400   REM - SUBROUTINE CALCULATES FFT OR INVERSE FFT
410   REM - N = # OF POINTS IN WAVEFORM (POWER OF 2)
420   REM - CODE = 1 FOR FFT, -1 FOR IFFT
430   REM - DELTA = dT FOR FFT OR 1/dT FOR IFFT
440   REM - R(N) = REAL DATA ARRAY FOR INPUT & OUTPUT
450   REM - I(N) = IMAGINARY DATA ARRAY FOR INPUT & OUTPUT
460   IR = 0
470   N1 = N
480   N2 = INT(N1/2)      'CHECK IF N IS A POWER OF 2
490   IF N2*2 < > N1 THEN PRINT "N IS NOT A POWER OF 2!": RETURN
500   IR = IR + 1
510   N1 = N2
520   IF N1 > 1 THEN GOTO 480
530   PN= 2! * PI/N
540   L = INT(N/2)
550   IR1 = IR -1
560   K1 = 0
570   FOR Z = 1 TO IR
580   FOR J = 1 TO L
590   K = K1 + 1
600   P = K + L
610   KAY = INT(K1/(2^IR1))
620   GOSUB 1030                    'BIT REVERSAL SUBROUTINE
630   AM = KBITR
640   IF AM < > 0 THEN GOTO 680
650   XR1 = R(P)
660   XI1 = I(P)
670   GOTO 730
680   ARG = AM * PN
690   C = COS(ARG)
700   S = -1 * CODE * SIN(ARG)
710   XR1 = C * R(P) - S * I(P)
720   XI1 = C * I(P) + S * R(P)
730   R(P) = R(K) - XR1
740   I(P) = I(K) - XI1
750   R(K) = R(K) + XR1
760   I(K) = I(K) + XI1
```

```
770   K1 = K1 + 1
780   NEXT J
790   K1 = K1 + L
800   IF K1 < N THEN GOTO 580
810   K1 = 0
820   IR1 = IR1 - 1
830   L = INT(L/2)
840   NEXT Z
850   FOR K = 1 TO N
860   KAY = K -1
870   GOSUB 1030              'BIT REVERSAL SUBROUTINE
880   K1 = KBITR + 1
890   IF K1 < = K THEN GOTO 960
900   XR1 = R(K)
910   XI1 = I(K)
920   R(K) = R(K1)
930   I(K) = I(K1)
940   R(K1) = XR1
950   I(K1) = XI1
960   NEXT K
970   IF DELTA = 1 THEN RETURN
980   FOR K = 1 TO N          'SCALE OUTPUT DATA BY DELTA
990   R(K) = DELTA * R(K)
1000  I(K) = DELTA * I(K)
1010  NEXT K
1020  RETURN
1030  REM -BIT REVERSAL SUBROUTINE
1040  REM - KAY = INPUT NUMBER
1050  REM - IR = NUMBER OF BITS TO REVERSE
1060  REM - KBITR = REVERSED NUMBER
1070  KBITR = 0
1080  KAY1 = KAY
1090  FOR Y = 1 TO IR
1100  KAY2 = INT(KAY1/2)
1110  KBITR = 2 * KBITR + KAY1 - 2 * KAY2
1120  KAY1 = KAY2
1130  NEXT Y
1140  RETURN
```

10.2.5 Convolutions and Other Transforms

The utility of FFTs extends far beyond simple frequency analysis of acquired signals, even though this is still an important application. In the real world it is often difficult to measure a quantity "cleanly," without distortion due to the measurement system itself. For time-based or distance-based measurements, the overall system response is a function of the measured quantity along with a function of the system response. This system-response transfer function operates on the desired physical quantity through a process called convolution, producing the measured response.

The convolution $h(x)$ of two time (or space) domain functions $f(x)$ and $g(x)$ is defined as

$$h(x) = f(x) \cdot g(x) = \int_{-\infty}^{+\infty} f(X)g(x - X)\, dX$$

We will use the symbol · here to denote convolution. Convolution literally means "folding back." The value of one function at a particular point (x value) affects the overall response at neighboring points, as shown by the $g(x - X)$ function. Convolving two transfer functions produces the overall system-response transfer function.

The convolution integral can be difficult to calculate in the time (or space) domain for many functions. It becomes a simple problem in the frequency domain. The convolution of two signals in the time (or space) domain is equivalent to multiplying their FFTs in the frequency domain. If the FFT of functions $f(x)$, $g(x)$, and $h(x)$ are respectively $F(y)$, $G(y)$, and $H(y)$:

$$H(y) = F(y) \times G(y)$$

where $h(x)$ is calculated from the inverse FFT of $H(y)$. This is illustrated graphically in Figure 10-9. Notice that once the FFTs are multiplied (point by point), an inverse FFT (IFFT) is performed on the result to produce the output impulse response, which is the convolution of the two input responses.

An important aspect of transforming convolutions into multiplications via FFTs is that we can reverse the process. If we have data acquired from a system with a known impulse response, we can correct for that impulse response. We transform the measured data, along with the impulse response, to the frequency domain (via an FFT). By dividing the FFT of the measured data by the FFT of the impulse response, we deconvolve the data. Transforming the result via an IFFT results in data fully corrected for the system's impulse response. This process is shown graphically in Figure 10-10.

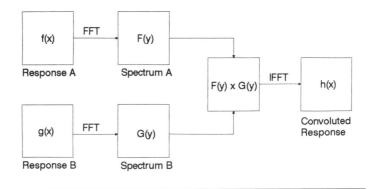

Figure 10-9 Convolution algorithm using FFTs.

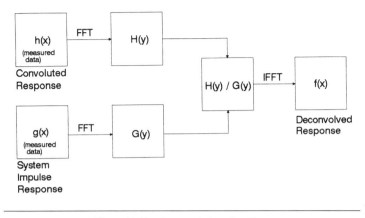

Figure 10-10 Deconvolution algorithm.

Deconvolution is an extremely useful analysis technique. In the field of optics, for example, image enhancements can be implemented via deconvolution. A simple example is a pinhole camera. An ideal pinhole, with a diameter much smaller than the wavelength of light, acts like a lens, producing an inverted image of an object, as shown in Figure 10-11a. Each point of the image corresponds to light from only a single point of the object. With a non-ideal pinhole, each image point corresponds to several object points, as in Figure 10-11b. The image becomes blurred as light from neighboring points mix together. This is the convolution of the real image with the light distribution function of the pinhole. Knowing that pinhole transfer function, we can deconvolve the data to get the undistorted image.

There are many other examples of the utility of deconvolution, as in the field of ultrasonics. Figure 10-12 shows a simple experiment using a pair of ultrasonic transducers in a water bath. An ultrasonic pulse is transmitted by one transducer and received by another transducer for data acquisition. The ultrasonic properties of the test sample, between the two transducers, is of interest. By deconvolving the measurement taken when the test sample is present with a measurement taken without the test sample, the impulse (and frequency) response of the entire test system can be eliminated from the data. This leaves the true ultrasonic response of the test sample. The test-sample frequency response provides information on its physical properties.

Another important DSP technique is digital filtering. One approach to filtering waveforms is to use FFTs to convert them into the frequency domain. Then an appropriate filter function is multiplied by the transform

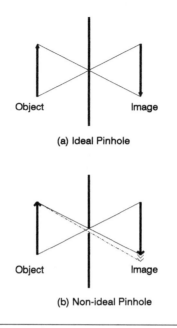

Figure 10-11 Optical pinhole.

of the input signal. Finally, the result is passed through an inverse FFT to produce the filtered time-domain waveform. For example, in the frequency domain, a band-pass filter can be just a simple rectangular function, set to zero outside the pass band.

When analyzing real-world data, there are often artifacts we wish to ignore. With ultrasonic measurements, for example, there are often pulse echoes. If we need to analyze the data of interest without including the entire waveform, often a windowing function is used. The simplest time-domain window function is a rectangular pulse, which is multiplied with the time-domain waveform of interest. The width and position of the pulse is selected so that it has a value of 1 over the region of interest in the waveform and a value of zero elsewhere, as illustrated in Figure 10-13.

Multiplying two functions (signal and window) in the time domain is equivalent to convolving their FFTs in the frequency domain. As we previously saw in Figure 10-6, the FFT of a rectangular function produces multiple peaks following the first main peak at zero frequency. These secondary peaks are referred to as side lobes. The higher the amplitude of the side lobes, the more the windowing function distorts the signal when they are transformed. For a rectangular window, the first side lobe has a peak amplitude of only -13 dB relative to the main (zero-frequency) peak.

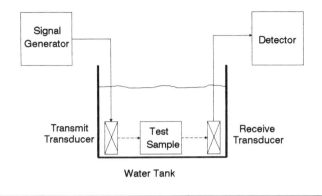

Figure 10-12 Simple ultrasonic test system.

Due to the convolution distortion, time-domain window functions other than simple rectangles are used. Several are based on cosine functions, which slowly taper to zero near the edges of the window region. Besides having lower side lobes, these windows also have wider main lobes than a rectangular function. This further helps to decrease any distortion they cause.

Two commonly used window functions are the Hanning and Hamming windows, shown in Figure 10-14. These window functions are defined for a width of N points, as follows:

$w(x) = 0.5 * \{1 - \cos[2\pi x/(N - 1)]\}$ Hanning window

$w(x) = 0.54 - 0.46 * \cos[2\pi x/(N - 1)]$ Hamming window

where x varies from point 0 to point $N - 1$.

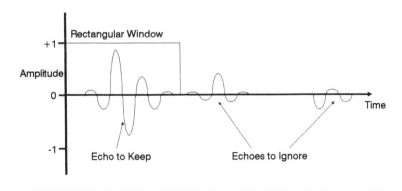

Figure 10-13 Using a rectangular window on ultrasonic echo waveforms.

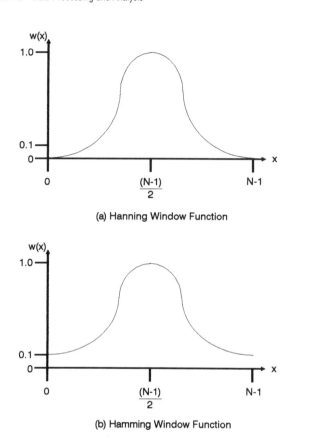

Figure 10-14 Hanning and Hamming window functions.

Notice that both window functions have their amplitude peak of 1.0 at the center of their range, $(N - 1)/2$. The main difference between them is that the Hanning window goes to zero amplitude at the edges of its range ($x = 0$ and $x = N - 1$), and the Hamming window has a finite amplitude of 0.08 at these edges. Both of these windows have a main lobe twice as wide as an equivalent rectangular window, with the same value of N. The Hanning window has a peak side lobe amplitude of -31 dB, and the Hamming window has a peak side lobe amplitude of -41 dB. These indicate a large improvement (18 to 28 dB) over the rectangular window's peak side lobe amplitude of only -13 dB.

Many other transforms are used for DSP analyses. One of these is the Hilbert transform. This is a technique used to obtain the minimum-phase response from a spectral analysis. When performing a conventional FFT, any signal energy occurring after time $t = 0$ will produce a linear delay component in the phase of the FFT. Even if a pulse occurs at $t = 0$, if it has finite width it will produce this linear slope in the resulting FFT phase. The slope of the FFT phase (versus frequency) is proportional to this time delay term. Significant delays can produce phase variations greater than 2π. If the FFT data contains phase nonlinearities of interest, they can be hidden by this large linear-phase component.

The Hilbert transform, based on special processing of an FFT, will produce a frequency response with this linear-phase component removed. This is the "minimum-phase" data desired. The algorithm involves signal processing in both the time and frequency domains.

10.2.6 Other DSP Techniques

A host of DSP techniques besides the FFT are commonly used. An exhaustive survey of the DSP field is outside the scope of this book. We will just look at a few more techniques that you may likely need in a data acquisition system. Please refer to the bibliography for sources of more detailed information on DSP.

Digital filtering techniques are often applied to time-domain signals, as in real-time filtering applications. Depending on system parameters, these digital filters can operate more quickly than the two FFT operations (forward and inverse) required for filtering a time-domain signal in the frequency domain. The two common types of digital filter approaches are finite impulse response (FIR) and infinite impulse response (IIR). The filtering process is effectively a convolution of the time-domain signal with a filter function.

FIR digital filters are considered nonrecursive. They mix delayed portions of the input signal with feedforward of the undelayed signal. They operate only on a small time-domain window of signal data. The filter function describes the coefficients for each of the delayed and undelayed components. FIR filters usually have a linear phase response, are relatively easy to implement, and do not tend to accumulate errors, since they operate on a data window of finite width. Their main limitation is the need to use many coefficients for good performance. This results in longer computation times and lower bandwidths.

IIR digital filters are considered recursive. They mix the input signal with time-delayed feedback of the output signal. They operate on a wide time-domain window of signal data. Even though it may be more difficult to design an IIR filter than a FIR filter, the resulting IIR filter is simpler, with fewer coefficients. This results in shorter computation time and wider bandwidths. Their main drawbacks are their sensitivity to noise and error accumulations, due to including effects of all past data.

The final DSP technique we will touch on here is cross correlation. This is used to see how similar two functions are. The cross correlation function of $x(t)$ and $y(t)$ is

$$c(t) = \frac{1}{(a_x a_y)} \int_{-\infty}^{+\infty} x(T)y(t + T)\, dT$$

where a_x and a_y are the RMS values of the two functions. This normalizes $c(t)$ to a maximum value of 1 (if the two functions correlate). If the two signals are very similar, there will be a maximum in the cross-correlation function. Otherwise, there will not be any significant maximum. If one function represents a delayed version of the other function, $c(t)$ will equal 1 at a value of t equal to the time delay.

This concludes our overview of data processing with personal computers. The techniques covered include some of the most common data analysis methods used with data acquisition systems. In the next chapter we will look at commercial hardware and software data acquisition products for personal computers.

CHAPTER

Commercial Data Acquisition Products

There is a plethora of commercially available data acquisition products for personal computers, with the number growing larger every day. The largest selection exists for MS-DOS PC/XT/AT systems. However, a growing number of products are now available for Apple's Macintosh and IBM's PS/2 personal computer lines. These commercial products fall into two broad categories: hardware and software. Software is usually included with most hardware products, to assist the user. Occasionally, hardware manufacturers just recommend compatible software products, along with programming guidelines. Some products are a complete hardware/software bundle, requiring both for proper operation.

In this chapter we will survey the vast array of data acquisition hardware and software products. We will look at products from a few major manufacturers in detail, including operational details as well as how to use the products. The Appendixes contain lists of commercial data acquisition product manufacturers (hardware and software). Since most hardware products operate similarly, regardless of the personal computer platform used (PC/XT/AT, PS/2, Macintosh), an in-depth discussion will again center on MS-DOS PC/XT/AT products. First, we will examine hardware products.

11.1 Commercial Data Acquisition Hardware Products

A large number of manufacturers produce data acquisition hardware products for personal computers. The largest market is for MS-DOS PC/XT/AT platforms. For these machines, most data acquisition hardware products are cards that plug into a PC's expansion bus. These cards come in two major versions: 8-bit cards for PC/XT class computers and 16-bit cards for AT (ISA) machines. Many products have additional hardware, external to the PC, which connect to the main data acquisition card. These add-on devices include connection boxes, signal conditioning boards, and high power I/O interfaces, including relay boards. Some PC-based data acquisition systems consist of an external box, connected to the PC's bus for control, usually via a special interface card.

Many data acquisition boards for PCs have dedicated functionality, such as analog inputs only. Some may have expansion capability, such as an additional multiplexer for more analog inputs. Other PC-based data acquisition cards are designed to be modular. They consist of a basic plug-in card, the *carrier*, which accepts several modules riding "piggy-back" on it. These modules offer specific functions, allowing the user to tailor the hardware to his particular needs (such as the number of analog inputs and outputs required). The module functions include analog I/O, digital I/O, and signal conditioning. This modular approach offers great flexibility, at a higher price. It is usually justified when a highly customized system is required or configuration changes will occur often.

Data communications interface cards are also important pieces of data acquisition system hardware. In this case, the personal computer is used as an intelligent controller, running remote data acquisition equipment through the interface. These interfaces include GPIB, RS-232C, RS-422, and RS-485. Of course, these cards can also be used in PCs for communications purposes other than data acquisition. For example, even though a GPIB interface card is often used in a PC to control automated instruments, it could be used to simply drive a printer or plotter.

Data acquisition cards for PCs fall into several major functional categories, including digital I/O, analog I/O, and counter/timer. Some boards have most or all of these features, others have only one or a few. A few typical PC/XT/AT data acquisition cards are shown in Figure 11-1. There are also specialized data acquisition cards which have features geared to a particular application, such as chromatography equipment used in analytical chemistry labs.

Digital I/O cards have input and output lines operating at TTL logic levels (in the range of 0 to +5 volts). Stand-alone digital I/O cards often contain some multiple of eight I/O lines, with 16 or 24 being most common. These cards can be used as parallel, digital interfaces as well as dedicated controllers. Most digital I/O cards allow programming lines for input, output, or both. Usually they contain interrupt-generating hardware. Some digital I/O cards support DMA for maximum data transfer speeds.

Figure 11-1 Typical single-function data acquisition cards for PCs. (Courtesy of Burr-Brown/Intelligent Instrumentation)

A popular IC used for digital I/O is the Intel 8255A Programmable Peripheral Interface (PPI), whose block diagram is shown in Figure 11-2. This device has three 8-bit ports that can be programmed for one of three modes: simple, unidirectional I/O without handshaking; strobed, unidirectional I/O with handshaking; and strobed, bidirectional I/O on the same pins, with handshaking. The 8255A is controlled by addressing its control port and three data ports. It is so popular that many board manufacturers provide 8255A software compatibility for digital I/O ports that do not use the 8255A.

Analog I/O cards are the most common form of data acquisition hardware for PCs. They contain one or more ADCs for analog input and DACs for analog output, either on the same or separate cards. Usually, any card containing an ADC for analog input is considered a data acquisition card. Analog input cards typically contain one ADC IC or module along with one or more analog multiplexers. This enables several analog signal sources (such as conditioned sensors) to be connected to one board at the same time. For example, multiple temperature sensors may be used in monitoring different portions of a piece of equipment under test. The multiplexer allows one of several analog inputs to be connected to the

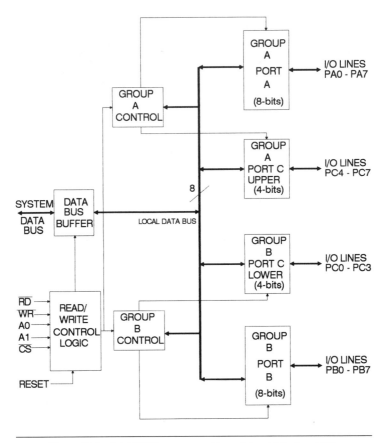

Figure 11-2 Intel 8255A programmable peripheral interface (PPI).

ADC at any given time. Commonly, commercial ADC cards have 8–32 analog input channels. These channels may be differential or just single ended.

The resolution of the ADCs and DACs used range from 8 to 16 bits. Analog I/O boards with 12-bit resolution are the most common, at present. Another important parameter is the maximum conversion rate for analog input cards. This can range from only tens of samples/second on high resolution and/or low-cost cards to over 100,000 samples/second on high-speed data acquisition cards with DMA hardware. Cards with conversion rates as high as one million samples/second are also becoming common.

When looking at maximum conversion rates for these ADC cards, remember that the rate is usually specified for a single channel only. If you need to measure several inputs simultaneously, the maximum conversion rate at any channel is the ADC's maximum rate divided by the number of multiplexed channels used. If this overall rate is too slow, you will need either multiple ADCs (one or more cards) or a faster ADC.

Analog output cards usually contain one DAC per output channel. Occasionally, a card may contain one DAC and several analog output channels, employing a sample-and-hold (S&H) amplifier for each channel. As we previously saw (in Chapter 3), S&H amplifiers "remember" a voltage level using a charged capacitor. Since the capacitor's charge slowly drops (due to its own leakage current and that of the surrounding circuitry), the S&H output "droops" with time. The S&H output must be continuously refreshed by recharging the capacitor (as with DRAMs), or the analog output will be valid only for a short period of time (usually on the order of milliseconds). Due to these drawbacks, this approach is not widely used. Most analog output boards have only a few channels, with an independent DAC for each.

Most analog I/O cards contain a timer/counter with multiple channels. This enables the card to perform conversions at a fixed rate, without any PC software overhead. It is a common option for data conversions to be controlled either by an internal (on board) clock, an external clock, or PC software commands. Most analog input cards have hardware interrupt capability. This is a programmable option, used to generate an interrupt when the ADC is ready to be read. It is especially important when the ADC conversion rate is controlled by a clock and is essentially asynchronous to the control software running on the PC.

Some analog I/O cards have DMA capability. This allows data to be transferred between the data acquisition card and the PC at the fastest possible rate. It does require special software support, but this is usually commercially available. These software packages are used for data *streaming* (transferring data between the data acquisition card and a disk file at high speed) as well as simulating the functions of a high-speed strip-chart recorder. Only high-speed analog I/O cards use DMA, since for slower cards the analog data conversion speed becomes the rate-limiting factor, not the data transfer rate.

Timer/counters are available on separate cards, typically in conjunction with digital I/O lines. Besides being used for controlling data conversion rates, they are also useful as general purpose clocks, frequency counters, and event counters. They usually have TTL compatible inputs, but with proper signal conditioning, such as an amplifier (to boost the signal level) and a comparator with hysteresis (to square up slow rise/fall times of a signal and convert it to TTL levels), analog signals can also be measured.

ICs commonly used for timer/counters are the Intel 8254 Programmable Interval Timer (PIT), as used in PC motherboards, and the AMD AM9513A System Timing Controller (STC), shown in the block diagrams of Figure 11-3. The Intel 8254 PIT contains three independent 16-bit counters, and the AMD AM9513A STC has five.

Each of the three Intel 8254 counters has a clock input, a gate input, and an output line. They are synchronous down counters (binary or BCD) with a count register to load a counter value, an output register to read the counter value, and a status register. Six programmable counting modes are available, allowing the 8254 to be used as a clock, an event counter, a one-shot generator, a programmable square-wave generator, or a complex digital waveform generator.

The AMD AM9513A is an extremely powerful counter/timer, with many operational modes and advanced features, making it a popular choice for manufacturers of high-performance data acquisition cards. Each of the five AM9513A counters has a source input, a gate input, and an output line. It differs from the 8254 by having a common clock generator on the chip itself. Each counter can choose its clock input from either this internal source (including a clock divided down from the internal one) or an external one, on its source line. This internal clock is typically 1 MHZ. The synchronous counters can count either up or down in binary or BCD. They can be concatenated for an effective counter length of 80 bits. The chip has a scaled frequency output. Each counter has a load register to initialize the counter, a hold register to read the instantaneous count value, and a mode register to program the counter's features (such as clock source, polarity of gating line, output conditions, etc.). The AM9513A can be used for extremely complex timing and waveform generation applications.

The most useful configuration for PC-based data acquisition hardware is the multifunction board. This card contains, at a minimum, an ADC and digital I/O lines. A typical multifunction data acquisition card contains several analog input channels, one or more analog output channels, several digital I/O lines, and several timer/counter channels. Some may even contain signal-conditioning circuitry, such as filters. These boards can contain all the hardware needed to convert a personal computer into a complete data acquisition system (along with the proper software), usually at a very attractive price. Just make sure that you need most of the functions on the card and each individual function meets your requirements (such as an adequate number of I/O channels or an ADC

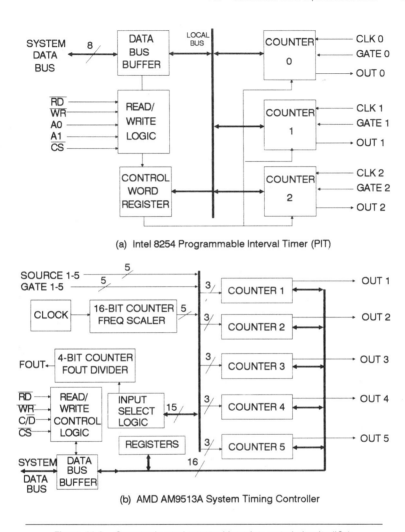

Figure 11-3 Commonly used counter/timer integrated circuits (ICs).

conversion rate that is fast enough). Without a doubt, multifunction boards are the most popular type of data acquisition card for a personal computer.

Now that we have covered some of the general aspects of data acquisition hardware, we will look at some commercially available products, covering a few of the more popular manufacturers. Complete addresses and other details are in Appendix A.

11.1.1 Keithley Metrabyte Corporation

This manufacturer produces data acquisition cards for PC/XT/AT computers as well as IBM PS/2 and Apple Macintosh systems. Keithley Metrabyte also produces PC-based industrial control and monitoring products, PC-based instrumentation products, and PC-based image processing products. Their products for PC-based data acquisition range from a low-cost analog input board, the DAS-4, to a high-performance, multifunction card, the DAS-20. The DAS-4 has eight single-ended analog input channels with 8-bit resolution and a maximum conversion rate of 3000 samples/second. It also has seven digital I/O lines. The DAS-20 has 16 single-ended (or eight differential) analog input channels of 12-bit resolution with a maximum conversion rate of 100,000 samples/second. The analog input range is software selectable. It also contains two 12-bit analog output channels with a maximum conversion rate of 130,000 samples/second. In addition, the DAS-20 has 16 digital I/O lines, two nondedicated counter/timer channels, interrupt support, and DMA support for both analog input and output. With additional accessories, it can support up to 128 analog input channels.

Keithley Metrabyte produces an extremely high-speed ADC board, the DAS-50, which has a maximum conversion rate of one million samples/second with 12-bit resolution. To support this data acquisition rate, which is faster than most PC DMA transfer rates, the DAS-50 has onboard memory, up to one million words (here one word = 12 bits), for data storage. It has four single-ended analog input channels, but no digital I/O or analog output lines. ADC triggering can come from a software command or an external logic or analog level signal.

Another high-performance Keithley Metrabyte product is the DAS-HRES, which has an ADC with 16-bit resolution, a maximum conversion rate of 47,600 samples/second and eight differential analog input channels. It also includes two 16-bit analog outputs, eight digital I/O lines and three counter/timers.

As an illustration of a typical data acquisition card for PCs, we will examine another high-performance Keithley Metrabyte board in greater detail, the DAS-16, shown in Figure 11-4. This is a multifunction card, with 16 single-ended (or eight differential) analog input channels of 12-bit resolution, with a maximum conversion rate of 50,000 samples/second (the DAS-16F, with DMA support, has a maximum rate of 100,000 samples/second). It has two 12-bit analog output channels, eight digital I/O lines, three timer/counter channels, and interrupt support.

The DAS-16 card will work in virtually all PC/XT/AT personal computers, as it requires only a PC/XT bus 62-pin expansion slot. A block diagram of the DAS-16 is shown in Figure 11-5. Like most PC-based data

11.1 Commercial Data Acquisition Hardware

Figure 11-4 Keithley Metrabyte DAS-16 multifunction data acquisition card. (Courtesy of Keithley Metrabyte)

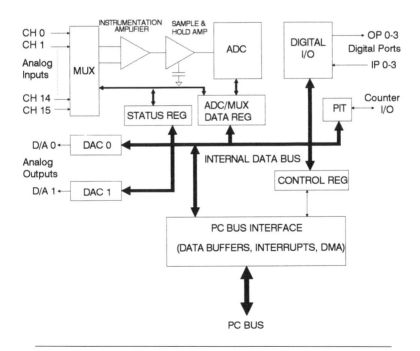

Figure 11-5 Block diagram of Keithley Metrabyte DAS-16 card.

acquisition cards, the I/O addresses used by the card are switch selectable. Since the DAS-16 uses 16 consecutive addresses, only the base or starting address is explicitly selected. By default this address is 300h, which is commonly used for data acquisition cards, being part of the I/O map (300h–31Fh) reserved by IBM for prototype cards. In this case, the DAS-16 would occupy 300h–30Fh. If this space was already in use, another base address would be selected, such as 310h. The base address has to fall on a 16-bit boundary, as the address select switches are for bits A4 through A9.

To aid in setting the base address, the DAS-16 comes with a program called INSTALL.EXE, on a utility disk. This program graphically displays the proper switch settings for the desired base address. In addition, it creates a file, DAS16.ADR, which contains this base address and is readable by application programs using the card. If the board address is ever changed, only the address in this one file needs to be changed to update multiple applications.

Other switches on the DAS-16 select differential or single-ended lines for the analog input channels, ADC gain level and unipolar versus bipolar ADC input range. The DAS-16 has five preset gain levels for the ADC, determining full-scale range. In bipolar mode these are ±10 V, ±5 V, ±2.5 V, ±1 V, and ±0.5 V. For some data acquisition cards, the gain levels are set by software commands.

All external connections to the DAS-16 (other than the PC-XT expansion bus connector) are made via a 37-pin D-shell connector at the back of the card. Most data acquisition cards use this type of arrangement if the number of connector lines is not excessive (usually 50 or less). The most common connectors used are D-shell and ribbon-cable varieties. If many external connections are needed, as with a multifunction card having a large number of analog and digital I/O lines, usually several ribbon-cable connectors on the board itself are used. These cables then have to be routed through an opening in the card's mounting bracket. On the DAS-16, the 37-pin D-shell connector contains all the analog and digital I/O lines. In addition, it contains control lines for the accessible timer/counters, power supply (+5 V) and reference voltage (−5 V) outputs, along with an input for an external DAC reference voltage (if a range other than 0 to +5 V is desired).

All software access to the DAS-16 is done by reading from and writing to the 16 I/O ports located between the base address and the base address plus 15. These I/O ports are listed in Table 11-1. Note that some of these ports are either read-only or write-only, while some are both read and write. In addition, the same port address can have a different function, depending on whether you read from it or write to it. For example,

TABLE 11-1
DAS-16 I/O Ports

PORT LOCATION	FUNCTION	READ/WRITE
Base Address + 0	ADC Low Byte Start ADC	R W
Base Address + 1	ADC High Byte	R
Base Address + 2	MUX Scan Control	R / W
Base Address + 3	Digital I/O Out (4 bits) Digital I/O In (4 bits)	W R
Base Address + 4	DAC 0 Low Byte	W
Base Address + 5	DAC 0 High Byte	W
Base Address + 6	DAC 1 Low Byte	W
Base Address + 7	DAC 1 High Byte	W
Base Address + 8	DAS-16 Status	R
Base Address + 9	DAS-16 Control	R / W
Base Address + 10	Counter Enable (2 bits)	W
Base Address + 11	Not Used	N / A
Base Address + 12	Counter 0	R / W
Base Address + 13	Counter 1	R / W
Base Address + 14	Counter 2	R / W
Base Address + 15	Counter Control	W

the base address, as a read port, returns the low byte of the last ADC conversion. As a write port it initiates an ADC conversion.

The mux scan port (base plus 2) allows multiple ADC channel conversions to be performed without explicitly stating the desired analog input channel prior to each conversion. The first and last channel numbers are written to this port. Each successive ADC trigger operates on the next analog input channel, within the range of first to last. After the last channel, the selection rolls around to the first one again. This feature is extremely handy if you are using multiple analog inputs with a hardware clock trigger. Once the software sets up the card to convert the desired ADC channels, all it has to do is keep reading the data until the required number of readings have been accumulated. Of course, the analog input channels used must be consecutive numbers.

The analog output ports (base plus 4 through base plus 7) are write-only, requiring two 8-bit ports to access the complete 12-bit DAC word. The DAC output is not changed until both bytes have been written, preventing a glitch in the DAC output when one byte is an old value and the other is a new one.

The eight digital I/O lines of the DAS-16 are configured as a 4-bit input port and a 4-bit output port. By writing to the digital I/O port (base plus 3), the four output lines are latched. Reading from the digital I/O port reflects the state of the four input lines. Two of the input lines are also used for special ADC trigger and counter gate functions.

The status port (base plus 8) is read-only. It contains information on the ADC and interrupt status. This information includes whether the ADC is busy or has valid data and if the analog inputs are single-ended or differential as well as unipolar or bipolar. This allows software to check the state of the hardware switches. In addition, the mux channel for the next conversion is noted here, along with the status of the board's interrupt generator.

The control port (base plus 9) is both a read and a write address and determines the operating modes of the DAS-16. It is used to enable or disable interrupt generation and select the hardware interrupt level to use (IRQ2–IRQ7). The control port can also enable DMA transfers (if enabled, the PC's DMA controller must be properly initialized). In addition, this port determines the source of the ADC conversion trigger: software only, internal timer control, or external trigger control.

The counter enable port (base plus 10), along with the four 8254 ports (base plus 12 through base plus 15), controls operation of the three counter/timer channels. Counters 1 and 2 are cascaded, so that counting periods ranging from microseconds to hours can be used to periodically trigger the ADC.

As with many other data acquisition board manufacturers, Keithley Metrabyte provides some software support with the DAS-16 for BASIC programs. This support includes BASIC examples in the manual, along with diskettes containing additional examples and an assembly-language driver, with routines callable from a BASIC program. This support includes both interpreted BASIC as well as compiled BASIC (see Chapter 13 for a discussion of programming languages). Drivers for other languages, such as C, Pascal, and FORTRAN, are also available from the manufacturer.

As an example, here is a small segment of a BASIC program that triggers an ADC conversion (via software) for a board at base address

BASADR, returns the 12-bit result in DAT, the analog input channel number in CHANL, and then displays the result:

```
10 BASADR = &H300                            'Default Base Address
20 OUT BASADR%,0                             'Start ADC conversion
30 IF INP(BASADR%+8)>= &H80 THEN GOTO 30     'Conversion Done?
40 LOW% = INP(BASADR%)                       'Read low byte
50 HI% = INP(BASADR%+1)                      'Read high byte
60 DAT% = 16 * HI% + INT(LOW% / 16)          '12-bit data read
70 CHANL% = LOW% AND &H0F                    'Analog channel number
80 PRINT "For Channel #";CHANL%;", ADC Value = ";DAT%
```

Note that the variable names end in % to signify they are integers (as opposed to floating-point numbers).

In Microsoft C, a similar program would look like:

```
#include <conio>          /* for inp() & outp() functions */
#include <stdio>          /* for printf() */
#define BASADR 0x300      /* default base address = 300h */
#define ADCLOW BASADR     /* address of ADC low byte */
#define ADCHI BASADR+1    /* address or ADC high byte */
#define ADCSTAT BASADR+8  /* address of status port */
main()
{                         /* start of program */
   int dat, low, high, chanl;  /* declare integers */
   outp(BASADR,0);        /* start conversion */
   while(inp(ADCSTAT)>=0x80);  /* wait for end of conversion */
   low = inp(ADCLOW);     /* read low byte */
   high = inp(ADCHI);     /* read high byte */
   dat = 16 * high + low / 16;  /* full 12-bit reading */
   chanl = low & 0x0f;    /* mask bits to get channel number */
   printf("\nFor Channel #%d, ADC Value = %d\n",chanl,dat);
}                         /* end of program */
```

This program may look more verbose than the BASIC version, but as the size and complexity of a program increases, the extra overhead of C is minimal compared to its flexibility, speed, and power.

This concludes our examination of the Keithley Metrabyte DAS-16 board. It is typical of many data acquisition cards for personal computers. We will continue now by looking at other leading manufacturers of data acquisition boards for personal computers.

11.1.2 Data Translation, Inc.

This manufacturer is another leading producer of data acquisition boards for personal computers. Their product line supports PC/XT/AT computers, IBM's PS/2 systems, Apple's Macintosh II computers, and other industrial computer platforms. They also produce software products for use with their data acquisition boards.

Data Translation's data acquisition card product line ranges from low-cost, single-function cards, such as the DT2814 analog input board (which will function in a PC/XT machine), to high-performance, multi-function cards, such as the DT2821 series (which require a PC with an AT or ISA bus). The DT2814 is just an analog input card, with 12-bit resolution and 16 single-ended input lines. It has a maximum conversion rate of 25,000 samples/second and analog input ranges of 0–5 V (unipolar), ± 2.5 V (bipolar) or ± 5 V (bipolar). It contains a programmable pacer clock, to initiate repeated conversions, and interrupt generation hardware.

The data acquisition cards in the DT2821 series contain either a 12-bit or 16-bit ADC with up to eight differential or 16 single-ended analog input channels. The maximum ADC conversion rate for the 12-bit boards ranges from 50,000 to 250,000 samples/second. For the 16-bit boards this maximum rate is either 30,000 or 100,000 samples/second. Most of the cards in this series have two analog output channels (DACs), with the same resolution as their ADC (12-bit or 16-bit). These analog outputs have maximum conversion rates of either 100,000 or 130,000 samples/second. The analog input amplitude ranges vary from ± 0.02 V to ± 10.00 V (bipolar) or 0–0.2 V to 0–10.00 V (unipolar), depending on the board. Analog output ranges can be fixed at ± 10 V (bipolar) or vary either from ± 2.5 V to ± 10.0 V (bipolar) or from 0–5 V to 0–10 V (unipolar). These boards all have 16 digital I/O lines, configured as two programmable 8-bit ports. They have a programmable pacer clock to initiate repeated data conversions, as well as interrupt and DMA support.

Data Translation has another interesting data acquisition product line, their DT2831 series. The boards in this series are very similar to their DT2821 counterparts with one important exception. Once the base address of a DT2831 card has been selected, all its data acquisition parameters are set by software only. There is no need to change switch or jumper settings to modify parameters such as analog input gain, single-ended versus differential analog inputs, analog voltage ranges, DMA channel, interrupt channel, or even ADC and DAC calibration. These boards support either 12-bit or 16-bit analog I/O. The maximum analog input conversion rate is 250,000 samples/second for 12-bit boards and 30,000 samples/second for 16-bit boards. The analog output conversion rates are 130,000 samples/second for 12-bit DACs and 100,000 samples/second for 16-bit DACs, with either a unipolar (0–10 V) or bipolar (± 10 V) output range. The DT2831 boards have eight digital I/O lines, configured as a single 8-bit port. They also have an AM9513A IC with five counter/timer channels. Three of these channels serve as internal pacer clocks for ADC and DAC conversions. The other two are available for external, general purpose uses. As implied above, these boards support hardware interrupts and DMA.

Data Translation, along with most other major manufacturers, produces screw-terminal and signal-conditioning panels for their data acquisition cards. These panels simplify connecting external devices to the data acquisition cards. Some common signal conditioning functions are available, such as programmable analog gain, and cold-junction compensation for thermocouples. If a thermocouple is directly connected to the appropriate panel, the analog signal sent to the data acquisition card can be directly read as degrees (temperature) without additional circuitry or complex software.

Note that some of Data Translation's cards will not operate in a PC/XT computer (containing an 8-bit data bus). They require an AT (ISA) personal computer (with a 16-bit data bus). It is becoming more prevalent for manufacturers to provide high-performance data acquisition cards only for ISA PCs. This is largely due to the higher performance inherent in an AT computer and the tendency of PC/XT computers to be considered obsolete, especially for laboratory and industrial applications.

11.1.3 Other Hardware Manufacturers

A large number of other PC-based data acquisition board manufacturers are listed in the appendixes. Without going into much detail, we will briefly look at a few more of them.

One manufacturer, Burr-Brown/Intelligent Instrumentation produces data acquisition products for PC/XT/AT, PS/2, and Macintosh II personal computers. Burr-Brown/Intelligent Instrumentation manufactures series of boards with dedicated functions as well as those with modular features, all part of their PCI-20000 system. This product line stresses the use of modular boards. Dedicated boards are available only for digital I/O, analog input, and analog output (as shown in Figure 11-1). Termination and signal conditioning panels are also available, as are GPIB interface cards.

The carrier boards used with expansion modules act as multifunction cards, plugging into a PC's I/O expansion slot. Some carrier boards require modules for digital or analog I/O, such as the PCI-20001C series. Others, as the PCI-20041C series, contain digital I/O but need additional modules for analog I/O. The PCI-20098C is considered a multifunction carrier board, containing analog I/O, digital I/O, and counter/timers, in addition to supporting additional modules.

The add-in modules for these carrier boards include various analog input options, such as high gain (up to ± 5 mV, full scale), high speed (up to 180,000 samples/second at 12-bit resolution), and analog input expansion. The analog output modules offer 12-bit or 16-bit resolution, with maximum conversion rates of 80,000 samples/second. A digital I/O module offers 32 lines, accessible as four 8-bit ports. Other modules with special functions include a counter/timer board, a sample-and-hold board, and a trigger/alarm board.

One very interesting product in this Burr-Brown PCI product line is their PCI-20202C series of smart carriers. These carrier boards are based on Texas Instrument's TMS320C25 DSP IC. Besides having conventional analog and digital I/O capability (with appropriate modules added), these boards can perform high-speed computations, such as spectral analysis and digital filtering, without burdening the PC. When used with compatible software (such as products available from Burr-Brown/Intelligent Instrumentation), it forms an easy-to-use DSP workstation (DSP ICs will be discussed further in Chapter 12).

Another major data acquisition board manufacturer is Analog Devices. Their RTI product line supports several computer platforms, including PC/XT/AT and PS/2 personal computers as well as other industrial computer systems. Their RTI-800 series of dedicated function boards are for PC/XT/AT systems. Most RTI-800 boards will work in a PC/XT or AT system, while some require an AT bus.

This RTI-800 series includes single-function and multifunction boards. The RTI-800 board has 12-bit resolution analog input (with 16 single-ended or eight differential input channels) at conversion rates up to 91,000 samples/second. It also has 16 digital I/O lines (as eight inputs and eight outputs) and three counter/timers. The RTI-802 has either four or eight analog outputs at 12-bit resolution. Its DACs have a 20-μsec settling time, limiting their conversion rate to less than 50,000 samples/second. The RTI-817 is a digital I/O board with 24 lines, configured as three 8-bit ports. It also has hardware interrupt capability.

The RTI-815 is a complete multifunction board. It has all the features of an RTI-800 board (12-bit analog input up to 91,000 samples/second, 16 digital I/O lines, and three counter/timers) along with two 12-bit analog output channels.

Analog Devices also offers high-performance cards in this series. The RTI-870 analog input board provides the extremely high resolution of 22-bits, at low conversion rates (up to 20 samples/second). It has four differential inputs and two 4-bit digital I/O ports. The RTI-860 boards offer extremely high-speed analog inputs, operating at 250,000 samples/second for 12-bit resolution or 330,000 samples/second for 8-bit resolution. These boards, which have local RAM to support the high data rates, require an AT bus. As other manufacturers, Analog Devices offers termination and signal-conditioning panels for its products.

An additional major manufacturer of data acquisition hardware for personal computers is Keithley Instruments, now the parent company of

Keithley Metrabyte and Keithley Asyst Technologies (see below). Keithley's approach to PC-based data acquisition systems is different than the manufacturers we have previously discussed. Instead of having the data acquisition hardware plug into the PC's bus, Keithley produces a stand-alone expansion chassis, which connects to the PC bus via a cable and interface card. The expansion chassis contains the data acquisition hardware, as selectable modules, while the overall system is still controlled by the PC.

Keithley's 500-Series data acquisition systems use this approach, working with a PC/XT/AT personal computer. The obvious advantages with this scheme are flexibility and expandability. In addition, signal conditioning can be added without using external components, and noise levels can be more easily controlled than inside a PC chassis. The disadvantages are a higher price and the extra space needed for an external chassis.

One of the first manufacturers of data acquisition boards for PCs was Scientific Solutions, Inc. Their product line supports PC/XT/AT and PS/2 personal computers. It includes multifunction data acquisition boards, digital I/O boards, and GPIB interface cards.

Scientific Solutions' Lab Tender board is a low-cost, multifunction card. It contains an 8-bit ADC with 32 single-ended or 16 differential inputs, having a range of ±5 V and a maximum conversion rate of 50,000 samples/second. The Lab Tender has an 8-bit DAC, multiplexed with 16 sample-and-hold outputs. If more than one output at a time is in use, they must be refreshed every 250 msec or less. This board has 24 digital I/O lines, configured as two 8-bit and two 4-bit ports, via an Intel 8255A IC. It also has five counter/timer channels, via an AMD AM9513A IC. The Lab Tender supports hardware interrupts.

Scientific Solutions also produces the multifunction Lab Master DMA board. This is effectively a greatly enhanced Lab Tender. The Lab Master contains a 12-bit ADC with a maximum conversion rate of either 40,000 or 167,000 samples/second and 16 single-ended or eight differential analog inputs. The analog input range can be either unipolar or bipolar, and the gain can be externally adjusted. The Lab Master has two independent 12-bit DACs with five selectable output ranges and a maximum conversion rate of 200,000 samples/second. In addition it contains 24 digital I/O lines and five counter/timers, as the Lab Tender. The Lab Master supports hardware interrupts as well as DMA data transfers.

The final hardware manufacturer we will discuss here is National Instruments. They are a leader in producing GPIB interface products for a broad base of platforms, ranging from personal computers to minicomputers. They also manufacture data acquisition boards for personal computers, including MS-DOS PCs, IBM PS/2, and Apple Macintosh II systems.

National Instrument's MS-DOS PC data acquisition products require an AT system. They produce digital I/O cards with either 24 lines (AT-DIO-24) or 32 lines (AT-DIO-32), configured as three or four 8-bit ports. Both boards have hardware interrupt capability. The AT-DIO-32 also supports DMA transfers and has three counter channels.

National Instrument's multifunction card is the AT-MIO-16. It has 16 single-ended or eight differential analog input channels with 12-bit resolution and a conversion rate up to 100,000 samples/second. Analog input gain is software programmable. The AT-MIO-16 has two 12-bit analog output channels, eight digital I/O lines (configured as two 4-bit ports), and five counter/timers, using an AM9513A.

Thic concludes our overview of commercial data acquisition hardware for PCs. Please refer to the appendixes for a more comprehensive listing of manufacturers. Next we will look at commercial software products.

11.2 Commercial Data Acquisition Software Products

The availability of abundant, powerful, easy-to-use software is probably the strongest incentive to employ a personal computer as the platform for a data acquisition system. In many ways, the variations in data acquisition software products mirror hardware products. These software products vary from simple drivers, tied to a particular manufacturer's boards and dedicated to data collection tasks (analogous to single-function hardware), to complete data acquisition/analysis/display software packages, supporting a broad range of hardware products (analogous to multifunction hardware).

As we discussed previously, a software driver is a special program that acts as an interface to a particular hardware device. It is used in conjunction with other software—the controlling program. The driver handles all the low-level interfacing, such as reading from and writing to a data acquisition board's I/O ports. It presents higher-level commands to the controlling program, such as initiate an ADC conversion on the selected channel and return the result. The driver takes care of all the I/O port commands, simplifying the controlling program.

A driver for a data acquisition card is an aid to writing a program that uses the card. It is specific not only to the board it supports but also to the programming language it is used with. By itself, a driver is not a full-blown software package—it is only a tool used to create the complete program. That is why a manufacturer offers different drivers for different programming languages, such as BASIC, C, Pascal and FORTRAN. Commonly, a manufacturer supplies a driver to support BASIC or C with the data acquisition board. Other drivers are usually available at modest cost. Sometimes different drivers are needed for different compilar manufacturers supporting the same language, such as Microsoft C and Borland's Turbo C. See Chapter 13 for a discussion of programming languages.

Software drivers are usually supplied in one of two forms when supporting a compiled language: a memory-resident module or a linkable module. A memory-resident module is supplied as a small program (typically with a .COM filename extension). The program is run, making the driver memory-resident. Occasionally, the memory-resident driver is provided as a .SYS file and installed as a DOS device driver, via the CONFIG.SYS file. Once this driver is loaded into memory, another program can call its functions, using a software interrupt.

A driver in the form of a linkable module is usually supplied as an .OBJ file. The user-written control program makes calls to functions in this module. After this control program is compiled, it is linked with the driver file, producing the complete executable program, as an .EXE file.

Nearly all hardware manufacturers supply some software with their products, usually just drivers or subroutine libraries, for use in developing custom programs. At present, hardware manufacturers are starting to offer complete, menu-driven software packages for use with their boards. Examples of this include Data Translation's Global Lab and National Instrument's Lab Windows software packages.

Fortunately, you do not have to be a programmer to use most PC-based data acquisition systems. Many software manufacturers (and increasingly more hardware manufacturers) produce easy-to-use, menu-based, and graphics-based data acquisition and support programs. These software products may have one or more general functions: data acquisition, data analysis, and data display. Several high-end integrated software packages provide all three functions, in varying degrees.

An important tradeoff in all types of software packages is whether the user interface is menu or icon-based versus command-driven. In either menu (text display) or icon-based (graphics display) software, the user chooses among many options presented by the program. These types of programs are the easiest to learn and use. Their drawback is lack of flexibility. The only functions (or combination of functions) available are those built into the selection system. If you need to do something different, you are out of luck, unless the desired function is added as a new feature in a product upgrade.

218 CHAPTER 11 Commercial Data Acquisition Products

In contrast, command-driven software is harder to use and not as intuitive to learn. It does, however, offer the maximum flexibility. All available commands can be used with any possible combination of parameters. In addition, most command-driven software packages allow for some level of programming. This can be as simple as stringing a series of commands together, as with a macro, or as complex as a true program with looping, conditional execution, and a wide range of data types.

It is common for manufacturers of command-driven software to include a menu-based shell with their packages. This shell acts as an easily learned command interface, buffering the user from the command-driven language itself. Once someone becomes familiar with the system and "outgrows" the menu shell, they can directly use the command-driven interface, for more flexibility. This is usually the best of both worlds. The main drawbacks are slower performance and a larger program, due to the extra shell layer in the software.

Figure 11-6 shows a simple comparison of menu-driven versus command-based interfaces. To save a data array as a plot file, using the menu system (Figure 11-6a) you would choose selections from several overlapping windows and finally specify the output filename. In the command-driven interface (Figure 11-6b), a single one-line command performs the same operation.

Software with data acquisition functions is absolutely necessary for all data acquisition systems. This type of software enables the user to set up the data acquisition board's parameters, record the desired amount of data at a specified rate for a specified time, and store the resulting data in a

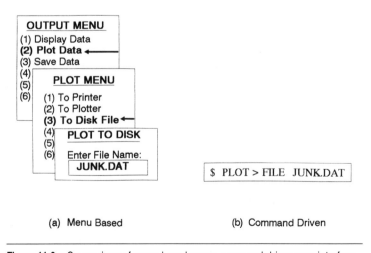

(a) Menu Based   (b) Command Driven

Figure 11-6 Comparison of menu-based versus command-driven user interface.

file, for future analysis and display. This software package must support the particular data acquisition board used as well as optional hardware accessories (such as multiplexers and signal conditioners).

This support takes many forms. There are many important factors to consider when comparing data acquisition software packages, especially which features are present and how they are implemented. Full support of all hardware features is mandatory, especially maximum ADC conversion rate, interrupt support, and DMA support. Support of multiple boards is certainly desirable. Some software packages include a real-time graphical display of the acquired data, in either an x-y graph or a strip-chart format. It is common for data acquisition software to scale the raw input values (usually signed integer format) from an analog input voltage to the physical units being measured by the sensors. For example, the millivolt readings from a thermocouple would be stored as degrees Celsius values, or the voltage output of a LVDT displacement sensor would be stored as millimeter values.

The format of the stored data (from an ADC) as well as how it gets stored are other important factors. Some software products will store data in ASCII format, which is easy to print out and read. ASCII data is readily transferred to other software packages for analysis and display. The drawback is that ASCII data takes up more disk space than equivalent binary data and requires more time to be stored and transferred. For relatively slow data rates or small amounts of data, this is not a significant problem. If the data rates get very high (DMA speeds, above 100K bytes/second, for example) or the produced data gets very large, a binary storage format should be used. In some cases, a data compression format may be needed, although this is not often done in real time. The binary format may be the same as the signed integer data produced by the data acquisition board, or it may be a more sophisticated format, including scaling and coordinate information, as when it represents a data array. Occasionally, a manufacturer may use a proprietary format. Usually, the data format is specified in the user's manual. This allows the user to transfer the data into another commercial or custom program.

Many standard data formats are supported by commercial data acquisition programs, besides ASCII. One of these is the Lotus 1-2-3 worksheet format. Spreadsheet programs, such as Lotus 1-2-3, are very popular for numerical analysis and display in business environments. Some data-acquisition-only software packages produce data outputs in this worksheet format, allowing the user to do analysis and output using Lotus or a compatible spreadsheet program. Other data acquisition software packages that provide analysis functions can both read and write data in this standard worksheet format.

One very specialized type of data acquisition software package is a data streamer. This is used to acquire data and store it in a hard disk file at the maximum rate allowed by the hardware. This maximum rate, sometimes referred to as throughput, is determined by the speed of the data acquisition board's ADC and the PC's maximum data-transfer rate to disk. If the data acquisition card has on-board memory, the PC's speed is not a factor, up to the board's storage capacity. A data streamer supports "no-frills," high-speed data acquisition. It is most useful with DMA hardware and provides little or no data processing.

Commercially available data analysis software is another important part of a PC-based data acquisition system. These analysis packages can be used by themselves, with data imported from other programs (data acquisition software, among others). Some of them also include data acquisition functions. The majority of data analysis software packages include output functions for CRT display as well as printer and plotter support. The capabilities of these packages vary from general mathematical operations to sophisticated DSP functions (such as FFTs). Some of these data analysis products are not even aimed at data acquisition—scientific or engineering applications. The business programs Lotus 1-2-3 and Symphony are examples. However, they are still very useful analysis tools for acquired data.

Some data-analysis software packages consist of function libraries for a particular programming language. They are analogous to hardware drivers, as they are only useful to someone creating a custom program. Of course, these libraries are not tied to any particular data-acquisition hardware. They may require a specific data format and interface only to particular language compilers. We will not dwell on these libraries here, as they are mostly of interest to software developers.

Software only for data display also exists, although most data analysis programs include some display capabilities. Data display usually consists of producing a graph or chart on a CRT screen, a printer, or a plotter. Useful data display features include ease of plotting data, changing scales, labeling plots, varying of plot formats (x,y line plots, bar charts, pie charts), using nonlinear axes (such as logarithmic), ability to output to a file, and support of a wide range of output devices (printers and plotters).

Now, we will examine a few data-acquisition software packages in greater depth, starting with multifunction programs.

11.2.1 ASYST

Keithley Asyst Software Technologies, Inc. produces ASYST, ASYSTANT, EASYEST, and ViewDAC. ASYST is a high-end data acquisition software product, specifically for MS-DOS PCs. It is a fully integrated programming environment with an extensive number of features for data acquisition, control, analysis, and display. It is a command-driven program, with an optional menu system (called Easy Coder) for novice users. ASYST is essentially a programming environment with the speed of a compiled language and the flexibility of an interpreted language (see Chapter 13 for a discussion of programming languages). You can execute commands, one at a time, or create an entire program to carry out a complex operation. ASYST requires a math coprocessor (Intel 80×87 IC) in the host PC to operate. (*Note:* This is an absolute requirement, not an option as in many other software packages.)

ASYST is based on the FORTH language. Each command or function is called a word. New words can be created by putting together existing words. This extensibility is extremely powerful, enabling an experienced user to produce very complex operations with a single command. Some of the ASYST words are for simple mathematical operations; others carry out sophisticated operations, such as FFTs. It is almost trivial to create a word in ASYST for deconvolution that would take the place of an entire, complex program written in a conventional programming language. The proper command syntax is very critical with ASYST, where even a blank space has a particular meaning (as a command delimiter). Learning to use ASYST fully is akin to learning a new language. Once it is mastered, its power and flexibility can be realized.

All operations in ASYST are stack oriented, employing "reverse-Polish" notation (as used on Hewlett Packard calculators). A stack, in computer terms, can be imagined as a stack of dishes in a cafeteria. As each new item is placed on the stack it pushes the entire stack down. As an item is taken off the stack, the entire stack pops up. Only the top of the stack is accessible. To operate on numerical quantities, they are pushed on the stack. The operation is performed, replacing the original quantities with the result, which can then be popped off the top of the stack.

For example, to add two numbers, each is pushed onto the stack by specifying each value, followed by the ⟨Enter⟩ key. The addition operation is then performed. The result at the top of the stack can then be displayed with line ? word (the question mark character), leaving the stack intact. The . word (the period character) displays the top of stack value and pops it off. The sequence of events to add 129.5 and 17.3 is as follows:

```
129.5 ⟨Enter⟩
17.3 ⟨Enter⟩
+
? ⟨Enter⟩
146.8
```

CHAPTER 11 Commercial Data Acquisition Products

A key feature of ASYST, for data analysis, is that the quantities placed on the stack for various operations do not have to be simple scalar numbers; they can just as easily be arrays or waveforms. Multiplying the values of two waveforms (of equal length) together is as easy as the scalar example above. Each waveform is placed on the stack, the multiplication word (*) is invoked, and the result is now on the top of the stack. Calculating the FFT of a signal is as simple as placing the waveform on the stack and invoking the FFT word.

ASYST contains a vast number of built-in functions. A recent version (3.10) contains over 1100 command words. The categories these words fall into include: basic mathematical functions, array operations, trigonometric functions, complex number operations, special analysis functions (such as curve fitting and FFTs), statistical functions, data acquisition hardware control, GPIB hardware control, RS-232 interface operations, file operations (including binary, ASCII, and Lotus 1-2-3 formats), interfacing to other languages (C and FORTRAN), DOS operations, graphics and plotting operations, string operations, and programming support functions. Figure 11-7 shows a typical ASYST graphics display screen.

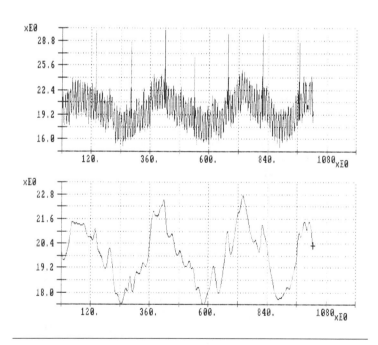

Figure 11-7   Sample ASYST display. (Courtesy of Keithley Asyst)

ASYST is sold as separate modules. The basic system, consisting of modules 1 and 2, contains analysis, statistics, graphics, and RS-232 interface functions. Module 3 adds support for data acquisition cards. Module 4 adds support for GPIB cards. It will operate on a PC/XT/AT or PS/2 system running DOS 2.0 or above, with a math coprocessor chip, a hard drive, and the full 640 Kbytes of RAM. It can use LIM expanded memory, if present. ASYST supports a wide range of printers and plotters. It also supports a wide range of data acquisition boards, including products from Analog Devices, Data Translation, Keithley, Keithley Metrabyte, and Scientific Solutions. It supports many different GPIB interface boards, including those from Keithley Metrabyte, National Instruments, Qua Tech, and Scientific Solutions, Inc. The program is copy-protected and comes with a hardware "key" that plugs into a PC's parallel port. It includes extensive documentation and 60 days of free support, with extended support as an option.

Keithley Asyst also produces an easy-to-use integrated software product, EASYEST. In effect, this is a scaled-down version of ASYST that is fully menu-driven and graphics-based. It is very easy for someone to learn how to use it, yet it remains a remarkably powerful analysis tool.

Another Keithley Asyst product is ViewDAC, a high-performance data acquisition, control, analysis, and display package, specifically for PCs using 80386 and 80486 CPUs. It is a menu-based system with programmability, which is multitasking. For example, it supports acquisition from up to 20 boards (with as many as 256 channels each) simultaneously.

11.2.2 LABTECH NOTEBOOK

Laboratory Technologies Corp. produces LABTECH NOTEBOOK, an integrated data-acquisition software package that runs on PC/XT/AT, PS/2, and Macintosh personal computers. LABTECH NOTEBOOK is a fully menu-driven program. It controls data acquisition from a wide range of boards, as well as performing some analysis and display functions.

Being menu-driven, LABTECH NOTEBOOK is extremely easy to use. You just keep selecting the appropriate menu and submenu entries, as well as filling in any necessary information. The options in the main menu are SETUP, GO, ANALYZE, CURVE-FIT, FFT, INSTALL, PROGRAM, and QUIT. For instance, SETUP configures many data acquisition and system control parameters. It must be used before any data can be acquired. The options in the SETUP submenu are CHANNELS, FILES, DISPLAY, VERIFY, and SAVE/RECALL. The CHANNELS option is used for data-acquisition parameter initialization. It, in turn, has a submenu consisting of NORMAL and HIGH-SPEED (the two acquisition modes). The submenus for each of these selections consist of numerous parameters to set (such as Sampling Rate, Duration, Trigger, Scale Factor).

LABTECH NOTEBOOK's functions can be automated using its special programming language, called MAGIC/L. This is an interpreted language, complete with math commands, looping, and control commands as well as LABTECH NOTEBOOK commands. Since it is an interpreted language (see Chapter 13 for a discussion of computer languages), its commands can be executed one at a time, acting as a command-based user interface.

Besides supporting a wide range of data acquisition boards (such as products from Burr-Brown/Intelligent Instrumentation, Data Translation, and Keithley Metrabyte), LABTECH NOTEBOOK has special support for certain common sensors: thermocouples, strain gages, and RTDs. This allows the software to compensate for sensor parameters up front, instead of during post-acquisition analysis.

One of the graphics features of LABTECH NOTEBOOK is displaying data as it is acquired. The program supports windows and customized displays. Data analysis is possible with LABTECH NOTEBOOK, but somewhat limited. The acquired data can be stored in either ASCII or binary formats. Another program, such as Lotus 1-2-3 or Symphony, is recommended for data analysis and printed output. Some data analysis can be performed within the program, as the data is acquired. The program also supports real-time open-loop or closed-loop process control. Figure 11-8 shows a sample LABTECH NOTEBOOK display screen.

LABTECH NOTEBOOK is first and foremost a data acquisition package. It combines power with ease of use. It will support a board's maximum data acquisition rate, in its HIGH-SPEED mode. It can also be used for control. It allows access to all the functions of a data acquisition card, including counter/timers. The MS-DOS version of LABTECH NOTEBOOK requires a PC/XT/AT or PS/2 system running DOS 2.0 or above, with a hard drive and a minimum of 512 Kbytes of RAM. It does not require a math coprocessor but will make use of one if present. It is copy protected and requires a hardware "key," which plugs into the PC's parallel port. Optional features for LABTECH NOTEBOOK include GPIB support, extended memory support, and real-time access, which provides real-time communications between the program and other software packages.

Laboratory Technologies produces another product, called LABTECH CONTROL. This is, effectively, a superset of LABTECH NOTEBOOK that adds industrial process monitoring and control functions. It

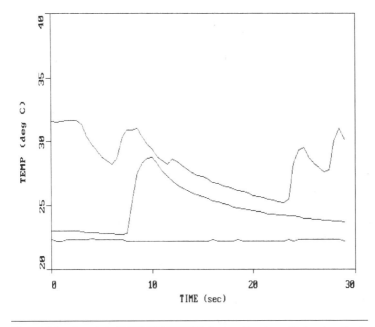

Figure 11-8 Sample LABTECH NOTEBOOK display. (Courtesy of Laboratory Technologies)

can display process flow diagrams with real-time data, using standard industrial symbols. It supports expanded memory and can run on a PC network.

Newer versions of LABTECH NOTEBOOK and LABTECH CONTROL also contain ICONview, a graphical interface. When used with a mouse, this software greatly simplifies the setup and data acquisition process.

11.2.3 DADiSP

DADiSP is a data analysis software package produced by DSP Development Corp. It is a menu-driven program that operates as an interactive graphics worksheet. It can display, process, and output data plots. It can handle up to 64 independent graphics windows simultaneously. Each window contains its own plot or data. Data in one window can be a function of data in other windows. As the independent data windows are changed, the dependent window is automatically updated. Suppose an independent window contains a waveform (perhaps acquired via another program). If a dependent window contains the power spectrum (via FFT) of that waveform, it will be updated if new data is read into the original waveform window. This is a powerful feature for doing complex analyses.

DADiSP contains over 260 functions (as of version 2.0). These functions include basic mathematical and trigonometric operations, statistical functions, FFT and related operations (such as autocorrelation), function generation, digital filtering functions, graphics operations, data file I/O, and hard-copy (plot or print) operations. It can optionally support GPIB data acquisition, via GPIB cards. Normally, its data input and output operate on files, using either ASCII or binary format. Data set size is limited only by disk storage space, not by available RAM. This enables very large waveforms to be manipulated intact.

DADiSP can directly call other programs, for uses such as data acquisition or specialized analysis. It can capture data from nearly any available source. It has advanced graphics features such as zooming and scrolling through a waveform. Many different plot formats are supported, including line graphs, histograms, bar charts, and scatter plots. You can annotate data windows with text or graphics, for labels and comments. It even supports transfer of its data to some desktop publishing packages.

A user can define custom menus in DADiSP, to help automate the analysis process and allow less skilled workers to perform it. The program also offers some programmability. It can monitor real-time processes and test for specified conditions.

DADiSP is supported on many different computer platforms, including MS-DOS PC/XT/AT and PS/2 systems. The basic version for PCs, DADiSP/PC, requires a PC/XT/AT or PS/2 computer with two floppy drives or a hard disk, 640 Kbytes of memory, MS-DOS 2.0 or later, and a compatible graphics card. A math coprocessor is optional but recommended. Currently, the most advanced version, DADiSP/V2.0, requires a PC with a 80286, 80386, or 80486 CPU (AT or PS/2) and 2 Mbytes of extended memory, along with DOS 3.0 or higher and a compatible graphics card. This version operates in protected mode.

11.2.4 Other Software Manufacturers

Many other software products are used in data acquisition systems, as listed in the Appendixes. Some of these software packages are produced by a hardware manufacturer for support of their boards only. Others are produced by independent manufacturers and support a wide range of hardware sources. Some analysis and display software packages are general purpose and not specifically for data acquisition.

SNAPSHOT STORAGE SCOPE from HEM Data Corp. is a menu-based data acquisition and display program that supports many hardware manufacturers (such as Burr-Brown/Intelligent Instrumentation and Keithley Metrabyte). It essentially treats the data acquisition system as a digital storage oscilloscope. Data is acquired, displayed, and stored as waveforms, as if using a storage oscilloscope. Up to 16 channels of data can be acquired simultaneously, with eight channels displayed at any given time.

Other menu-driven products from HEM add data analysis functions. SNAP-CALC performs a variety of mathematical operations on previously acquired data, including trigonometric and statistical functions. It supports the use of macros for frequently called operations. SNAP-FFT performs frequency analysis of data. It can produce output graphs with either linear or logarithmic scales.

MATLAB, from The Math Works Inc., is a data analysis and display program for use on a wide range of platforms, including MS-DOS (PC/XT/AT, PS/2) and Macintosh personal computers. It performs standard mathematical and statistical functions, with strong support of matrix operations. This program requires a math coprocessor IC, since all its calculations are done in floating-point format.

MATLAB can handle both real and complex data. It can input and output data files in several formats, including ASCII and binary. It has many graphics functions, including the capability of producing 3-dimensional plots. It produces high quality output plots on a wide range of printers and plotters.

MATLAB is an interactive, command-driven program. It is, effectively, an interpreted language (like BASIC), where an entire program can be run at once, or a single command at a time can be executed as it is entered by the user. Its integrated graphics capabilities make this program a powerful tool for scientific and engineering data analysis operating on both acquired and simulated data.

MATLAB's optional Signal Processing Toolbox adds many analysis functions useful in a data acquisition system. Some of these functions are for digital filter design (FIR and IIR), spectral analysis (including FFTs and Hilbert transforms), correlation, convolution, 2-dimensional signal processing (FFTs, correlation, and convolution), and windowing (including Blackman, Hamming, and Hanning windows).

The final software product we will examine here is an example of a data display and plotting program. TECH*GRAPH*PAD, from Binary Engineering, is a graphics program designed to display data and print or plot output graphs. This program is menu-based, making it easy to learn and use. It manipulates data files created internally or externally. Some of the data file formats supported by TECH*GRAPH*PAD are ASCII, binary, and Lotus 1-2-3 worksheet format. It can even import data directly stored by many acquisition cards (such as twos-complement binary data).

TECH*GRAPH*PAD has a file editor, allowing you to enter, edit, or manipulate data. It contains some data analysis capabilities, including curve fitting and data smoothing. It can also plot directly from equations, for dealing with theoretical data. This program supports several plot formats, including linear, log, and polar. Its graphics output can go directly to a printer or plotter or to a file for use with a word processing or desktop publishing package. It supports a wide range of printers and plotters.

This finishes our survey of some popular software products useful for data acquisition systems. We will end this chapter with a brief discussion of how to select the appropriate commercial hardware and software products for your data acquisition needs.

11.3 How to Choose Commercial Data Acquisition Products

Chances are, if you're putting together a PC-based data acquisition system, you will try to do it mostly with commercial products. This approach will not only save you a lot of development time but probably some money, also.

The first step in selecting your data acquisition system components is to define the physical measurements you need. For example, an environmental test chamber may require 10 temperature transducers covering the range of 0 to +150°C, with an accuracy of ±1°C and a reading from each transducer every second.

Next, determine the type of transducers to be used, the signal conditioning needed, and whether any output control signals are required. Continuing with the environmental test chamber example, since high accuracy is not required, thermocouples are a reasonable transducer choice. This would entail using a cold-junction compensation board to condition the thermocouple signals. Additionally, the ADC used would need moderately high gain, to deal with the millivolt level signals from the thermocouples. Let us also assume that one or more of the temperature sensors will control the temperature of the test chamber. This temperature control would be an analog signal, in the range of ±10 V, controlling the chamber's heating and cooling systems. So, we also need an analog output channel.

The following step is to consider how much data will be collected, how much analysis will be performed on it, and how it will be displayed.

In our example, we will assume the maximum test run time will be 1 hour. At a sampling rate of 1 sample/second per channel for 10 channels, 1 hour corresponds to 36,000 samples. For 8-bit data, this would produce a binary file of 36,000 bytes. For 12-bit data, a binary file would usually be 72,000 bytes long, if the data is not compressed (assume 2 bytes/sample). The data display should be all 10 temperature channels (transducers) in real-time with data saved to disk. The only analysis required would be the minimum, maximum, and average temperature of each channel.

The next step is to decide on the computer platform to use. Often, this is determined by the personal computers already on hand. If it is an open question, consider costs, which software packages you want to run on it, and whether this personal computer will be permanently dedicated to the data acquisition task or freed up at a later date for other jobs. Usually, software compatibility and availability is much more important than the computer's raw processing power, except for specific, high-performance applications.

The final step is picking out the hardware and software products to use. The most important factor here is hardware/software compatibility. Be absolutely sure that the software package you want will work properly with the hardware selected, including any signal conditioning boards, multiplexers, and other expansion modules.

For our test chamber example, high-performance is not required. The overall data rates are slow (10 samples/second). The required resolution could fit an 8-bit ADC, with 1-degree accuracy over a 150-degree dynamic range, fitting within 1 part in 256 (8 bits). However, low-speed 12-bit ADCs are not very expensive, and the extra resolution will produce better data. A 12-bit analog I/O board is a good choice, with at least 10 single-ended input channels and one output channel. At the low data-acquisition speeds called for, the board does not need a timer counter, since the PC's internal clock (normally 18 ticks/second) is adequate. A multifunction data-acquisition card would be overkill, unless the system may be used for other purposes in the future. In addition, a signal-conditioning panel with cold-junction compensation for thermocouples would be very useful.

Now that we have all our hardware specifications, we should pick out the software. We will assume that we do not want to write any programs and we would like the system to be operable by virtually anyone, without extensive training. This certainly points to a menu-driven data-acquisition software package. We do not need much analysis power, and we want a real-time display, with data storage and some means of producing a printed graph. In addition, automatic processing of thermocouple signals is required, to produce outputs directly in degrees Celsius. The software should also be capable of controlling the temperature of the test chamber. A good choice here would be a package similar to LABTECH NOTEBOOK.

This ends our overview of commercial data acquisition products for personal computers. This is a dynamic field, with new products (and manufacturers) appearing all the time. Please refer to the appendixes for more comprehensive listings of manufacturers and products.

In the next chapter, we will look at two personal computer architectures we have previously just touched on: IBM's PS/2 Micro Channel systems and Apple's Macintosh II systems. In addition, we will briefly examine specialized PC hardware products, such as math coprocessors and ruggedized PCs.

CHAPTER

Other Personal Computer Systems and Hardware

We have so far focused our attention on MS-DOS personal computers based on the IBM PC/XT and AT (ISA) buses. Several other types of popular personal computers are used for data acquisition systems. These include IBM's PS/2 computer line, based on their Micro Channel Architecture (MCA), and Apple's Macintosh II line, based on the NuBus interface. A third computer type is based on the Extended Industry Standard Architecture (EISA), a 32-bit extension of the 16-bit ISA bus, supported by several manufacturers. As of this writing, the EISA bus is not heavily supported yet and few data acquisition products are available for it.

We will examine the PS/2 MCA and Macintosh II NuBus systems in some detail. Then we will look at some additional PC hardware products, such as math coprocessors and specialized PCs.

12.1 IBM PS/2 Personal Computers with MCA

In 1987 IBM introduced its successor to its popular PC/XT/AT computer line, the PS/2. Most (but not all) members of the PS/2 family are based on IBM's Micro Channel architecture, including Models 50, 60, 70, and 80. Models 50 and 60 use Intel's 80286 CPU and Models 70 and 80 use the 80386 CPU. These MCA personal computers are not hardware compatible with the PC/XT/AT systems, but they are software compatible. They require very different expansion cards but will run MS-DOS (or OS/2) and nearly all PC application software. A data acquisition card for one of these systems must be built specifically for the MCA bus.

The performance characteristics of MCA are a large improvement over the ISA bus. The data bus is either 16 or 32 bits wide (depending on the computer model). On a 16-bit system, the address bus is 24 bits wide, for a physical address space of 16 Mbytes. On a 32-bit system, the address bus is also 32 bits wide, for a physical address space of 4 Gbytes. There are 11 hardware interrupt lines available (IRQ 3–7, IRQ 9–12, IRQ 14–15), which are level-sensitive. The same interrupt can be shared by multiple devices with open-collector drivers. DMA is supported with eight channels and a maximum transfer rate of 5 Mbytes/second (for 16-bit transfers). The maximum system data transfer rate with MCA is 20 Mbytes/second, for either memory or I/O cycles (under special conditions).

Figure 12-1 shows the 16-bit and 32-bit MCA connectors. Both are dual-row edge connectors with separate 8-bit and 16-bit signal sections.

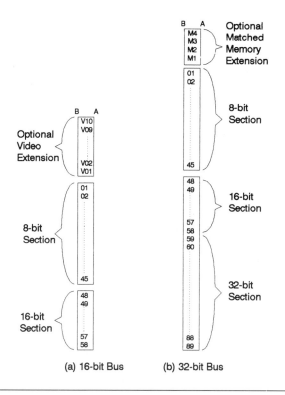

Figure 12-1 IBM Micro Channel bus connectors.

The 16-bit connector has 58 pins (each side) with 10 optional pins for video signals. The 32-bit connector, having a 32-bit section, contains 89 pins (each side) and has four optional pins for special memory transfer support. The MCA connectors place a ground pin between every four signal pins to minimize electromagnetic interference (EMI), which is a large improvement over the ISA connector layout. MCA also supports audio signals through a special pair of bus lines.

The signal assignments for a 16-bit MCA connector are shown in Figure 12-2. Most of these signals are similar to their ISA bus counterparts, such as address (A00–A23), data (D00–D15), interrupt (-IRQ03, -IRQ04, ..., -IRQ15), and control (-REFRESH, -TC, OSC) lines. Others are new to MCA, such as arbitration control lines (ARB/-GNT, ARB0–ARB3, -PREEMPT), connector-specific lines (-CD SFDBK, CD CHRDY, -CD SETUP), and other features (AUDIO).

IBM's Micro Channel is primarily an asynchronous bus. Data transfers over the Micro Channel are controlled by handshaking signals, instead of relying on a synchronous clock for transfer timing. When a device on the bus (such as an adapter card) is commanded to send or receive data, it responds with an acknowledge when it is done. These control and handshake signals determine the MCA timing. In addition, MCA does support certain synchronous data transfers.

MCA supports up to 15 bus masters in a system (including the main CPU), allowing multiprocessing with several CPUs, as well as DMA. A hardware-based arbitration scheme allows multiple bus master boards relatively fair access to the system bus. In addition, MCA systems have eliminated the need for setting switches or jumpers to configure a board (for addressing as well as interrupt and bus master arbitration levels) using its Programmable Option Select (POS) features. POS consists of registers that allow a board to be configured by software only. Each board manufactured for MCA receives a unique identification code from IBM, distinguishing it from other boards. A utility program automatically configures one or more boards in the system, assuring that addresses do not clash.

In spite of the major hardware differences between MCA-based PS/2 personal computers and PC/XT/AT systems, they are still software compatible. All PS/2 computers can run MS DOS and most of its available application software. Compatibility is assured if the software interfaces to hardware devices use BIOS or DOS function calls (as a "well-behaved" application). Any software that directly accesses hardware addresses (memory or I/O ports) may not be PS/2-compatible. One additional compatibility point is that all PS/2 systems use 3.5-inch diskette drives (either 720-Kbyte or 1.44-Mbyte capacity). An external 5.25-inch

234 CHAPTER 12 Other PC Systems and Hardware

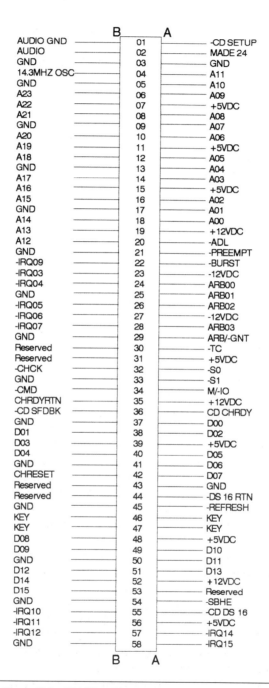

Figure 12-2 IBM Micro Channel 16-bit connector pinouts.

diskette drive is available as an option. Most commercial PC software packages are now available on 3.5-inch diskettes.

Many chip manufacturers support the MCA interface, simplifying the process of redesigning an ISA board to work in an MCA system. There are even commercially available ICs that directly convert MCA bus signals into an ISA interface, quickly changing an ISA board design to MCA. However, a redesign from the ground up is a much better approach.

The various MCA PS/2 systems have different processor speeds and amounts of motherboard memory. IBM's PS/2 Model 50, based on the 80286 CPU, has a CPU clock frequency of 10 MHz and supports up to 1 Mbyte of memory on the motherboard. As with all PS/2 systems using Micro Channel, memory can be expanded by adding boards to the channel, up to a system limit of 16 Mbytes. This limit also applies to the 80386-based models, with their 4-Gbyte physical address space (32 bits), due to the 24-bit address limitation of the DMA controller used in all these systems.

The PS/2 Model 60 is another 80286-based system, with a 10-MHz CPU clock and 1 Mbyte of motherboard RAM. The PS/2 Model 70 uses an 80386 CPU, running at 16 MHz, supporting up to 8 Mbytes of RAM on the motherboard. The PS/2 Model 80 system is also 80386-based, with a 20-MHz clock and up to 4 Mbytes of motherboard RAM. IBM is constantly adding new members to its PS/2 family as well as upgrading the performance of existing models.

Many data-acquisition hardware manufacturers who support PC/XT/AT systems also produce MCA products for PS/2 computers. This is true of most PC board manufacturers, in general. The listings of data-acquisition board manufacturers in the appendixes indicate PS/2 (MCA) support. Usually, an MCA board comes with a configuration program. Other than that, it can operate with other MS-DOS software packages. MCA boards are still not as plentiful as PC/XT/AT boards; but they are readily available, and their number is growing steadily.

Here is a small sampling of data acquisition boards for PS/2 MCA systems, available from major manufacturers. Keithley Metrabyte has several MCA products including a μCDAS-16G multifunction card, with 12-bit ADC resolution, up to 70,000 samples/second conversion rate, two 12-bit DACs, and eight digital I/O lines. This board is software compatible with their DAS-16G PC/XT/AT card. Keithley Metrabyte's μCDAS-8PGA is the Micro Channel equivalent of their DAS-8PGA, with 12-bit analog inputs at a maximum conversion rate of 20,000 samples/second. This analog input board, which is software compatible with the DAS-8PGA, also contains seven digital I/O lines. Other Keithley Metrabyte

MCA products include analog output, digital I/O, counter/timer, and communications boards (including GPIB).

Data Translation supports data acquisition on MCA systems with their DT2901 multifunction board. This card has 12-bit analog inputs with a maximum conversion rate of 50,000 samples/second. It has two 12-bit analog output channels along with 16 digital I/O lines. A related board is the DT2905, which is equivalent to the DT2901, except for higher analog input gain.

Another manufacturer supporting PS/2-based data acquisition is Analog Devices. Their RTI-204 multifunction board has 12-bit analog input channels with a maximum conversion rate of 19,000 samples/second, two counter/timers, two 12-bit analog output channels, and eight digital I/O lines. Other Analog Devices PS/2 products include high-density digital I/O and analog I/O expansion cards.

National Instruments also supports MCA, with its MC-MIO-16 multifunction board. This card has 12-bit analog inputs at a maximum conversion rate of 91,000 samples/second, two 12-bit analog outputs, three counter/timers, and eight digital I/O lines. Other National Instrument products for PS/2 systems include digital I/O and GPIB interface boards.

Burr-Brown/Intelligent Instrumentation is another manufacturer now producing data acquisition boards for MCA PS/2 systems. The PCI-601W and PCI-602W multifunction cards provide 12-bit analog inputs with conversion rates up to 70,000 samples/second, two 16-bit analog outputs (PCI-602W only), two counter/timers, and 16 digital I/O lines.

By way of comparison with PS/2 systems, the EISA bus, which was developed later as an alternative to MCA, is just starting to gain acceptance. It is an extension of the standard AT (ISA) bus from 16 to 32 data bits, along with 32 address bits. The EISA bus retains hardware compatibility with existing ISA boards. Of course, EISA cards must be used in an EISA computer to obtain the potential performance improvements, similar to MCA: high data-transfer rates (up to 33 Mbytes/second), multiple bus master support, automatic system configuration, slot-specific addressing.

Currently, the selection of EISA boards is much smaller than MCA boards. It is an open question whether EISA will become an established standard and receive widespread support. Its main advantage, at present, is being able to use standard ISA cards in an EISA personal computer, with the potential for obtaining higher performance using EISA cards. In the realm of data acquisition, National Instruments produces a board with a conversion rate of 1,000,000 samples/second for EISA systems (model EISA-A2000). Scientific Solutions produces the LabMaster Advanced

Design board, with 12-bit analog inputs and a maximum conversion rate of 333,000 samples/second, two 12-bit DACs, five counter timers, and 16 digital I/O lines. Bus data transfer rates for this board range from 350,000 to 500,000 words/second in an ISA bus and increase to the range of 1,000,000 to 2,000,000 words/second in an EISA bus (depending on the EISA mode). Not many other data-acquisition board manufacturers are actively supporting EISA systems yet.

This concludes our overview of IBM's PS/2 MCA systems. Next, we will look at Apple's Macintosh II personal computers, based on the NuBus.

12.2 Apple Macintosh II Computers with NuBus

In recent years, Apple's Macintosh computer line has gained acceptance for scientific, engineering, and industrial applications. This is due, in large part, to the open expansion architecture adopted for the Macintosh II series, based on the NuBus. The original Macintosh series, introduced in 1984, used a proprietary architecture and was not widely supported by third-party product manufacturers. This is in contrast to the open architecture of IBM's PC/XT/AT and PS/2 lines, as well as Apple's original computers. Many independent manufacturers produced hardware and software products for those systems.

Gradually, Apple opened up hardware accessibility to Macintosh computers, with a SCSI interface and a direct CPU interface, in some models. This permitted the use of a Macintosh with non-Apple hardware, including data acquisition products. The first truly expandable models were the Macintosh SE series. However, these units still had limitations. The Macintosh II series, based on the NuBus expansion bus, is a personal computer line with the expandability needed for serious engineering and scientific applications. These systems have between three and six NuBus expansion slots, allowing customization as readily as most ISA bus machines.

The Macintosh personal computer series is based on Motorola's M68000 microprocessor family. The 68000 CPU, used in the original Macintosh models (as well as the Macintosh Plus and Macintosh SE), has a 16-bit data bus and a 24-bit address bus, for a 16-Mbyte address range. The 68020 and 68030 CPUs, used in the Macintosh II series have a 32-bit data and address bus, for a 4-Gbyte address range.

The initial (and continuing) attraction of using a Macintosh computer is its graphics-based user interface. It has an intuitive operation, speeding up the learning process. The software burden rests with the program developers, not the users. A mouse is the standard user interface device with a Macintosh. The user interface is consistent across all Macintosh application software, minimizing the time needed to learn new programs. This type of graphics-based user interface is now available in the MS-DOS world, through software environments such as Microsoft Windows.

Because it is the most expandable Macintosh series and the one most widely supported by third-party hardware products, we will only consider Macintosh II systems here (with NuBus expansion slots) for data acquisition applications. The number of data-acquisition hardware manufacturers producing products for the Macintosh II implementation of the NuBus continues to grow steadily, along with application software.

NuBus is a system bus (as opposed to a CPU bus) developed by MIT and Texas Instruments for 32-bit computers. It is independent of the computer's CPU, providing buffered, multiplexed signals to the expansion connectors. A CPU bus is typically just an extension of the computer's internal CPU lines (with or without buffering) with some additional control signals added; it is tightly coupled to that CPU. As a system bus, none of the NuBus signals are processor-specific.

Apple's NuBus is a synchronous, multiplexed bus, using a 96-pin Eurocard DIN connector (popular in many industrial computer systems, such as those based on the VME bus). The form factor for NuBus cards (approximately 4.0 inches by 12.7 inches) is similar in size to PC/XT cards. It is based on the 1986 IEEE specification, IEEE-1196 NuBus (which originally called for a triple-height Eurocard form factor of 11.0 inches by 14.5 inches). NuBus is a synchronous bus (in contrast to IBM's asynchronous MCA), where all transactions are based on a fixed clock cycle. The edge of this clock determines the bus timing parameters, such as when data is valid or when it should be latched. Apple uses a 10-MHz bus clock in its NuBus. The address and data signals are multiplexed onto 32 lines (/AD0–/AD31). Various control signals are used to interpret these multiplexed lines. For example, the /START signal is asserted (active low) when the address/data lines contain a valid address.

The advantage of using synchronous bus transfers is simplicity of protocol and hardware to implement it. The advantage of using multiplexed address/data lines is the need for fewer bus wires (32 saved in this case) than in a nonmultiplexed arrangement. The disadvantage with this multiplexing is slower bus throughput, since multiple bus cycles are required for any data transfer (with separate bus cycles for sending address and data information).

The 96-pin NuBus DIN connector is arranged as three rows (A,B,C), of 32 pins, each. Due to address/data multiplexing, all the needed signals fit on 51 lines. The rest of the lines are used for power supply and ground, as shown in Table 12-1.

Each NuBus slot has its own unique ID number, with a maximum of 16 slots allowed by the specification, so each card knows the slot it is in. The 32-bit addressing space of the NuBus has a range of 4 Gbytes. The upper 256 Mbytes of this space is divided among the 16 possible slots, to provide each one with its own dedicated slot space of 16 Mbytes. Apple only uses up to six expansion slots in its Macintosh II systems, with slot IDs of 9 to Eh. Since each board knows the slot it occupies, it can automatically adjust its address mapping. This prevents address clashes

TABLE 12-1
NuBus Connector Pin Designations

Pin Number	Row A	Row B	Row C
01	-12VDC	-12VDC	/RESET
02	reserved	GND	reserved
03	/SPV	GND	+5VDC
04	/SP	+5VDC	+5VDC
05	/TM1	+5VDC	/TM0
06	/AD1	+5VDC	/AD0
07	/AD3	+5VDC	/AD2
08	/AD5	reserved	/AD4
09	/AD7	reserved	/AD6
10	/AD9	reserved	/AD8
11	/AD11	reserved	/AD10
12	/AD13	GND	/AD12
13	/AD15	GND	/AD14
14	/AD17	GND	/AD16
15	/AD19	GND	/AD18
16	/AD21	GND	/AD20
17	/AD23	GND	/AD22
18	/AD25	GND	/AD24
19	/AD27	GND	/AD26
20	/AD29	GND	/AD28
21	/AD31	GND	/AD30
22	GND	GND	GND
23	GND	GND	/PFW
24	/ARB1	reserved	/ARB0
25	/ARB3	reserved	/ARB2
26	/ID1	reserved	/ID0
27	/ID3	reserved	/ID2
28	/ACK	+5VDC	/START
29	+5VDC	+5VDC	+5VDC
30	/RQST	GND	+5VDC
31	/NMRQ	GND	GND
32	+12VDC	+12VDC	/CLK

with other boards and eliminates the need for configuration switches or jumpers (similar to IBM's POS in MCA systems). Since the Macintosh (and all M68000-family computers) uses memory mapping for I/O ports, this automatic configuration applies to both memory and I/O addresses.

Macintosh II systems access the NuBus in one of two modes: 24-bit and 32-bit. The "native" 32-bit mode allows full access to the bus and 16 Mbytes per slot. The 24-bit mode allows compatibility with earlier Macintosh systems, using an MC68000 CPU with 24-bit addressing. This 16-Mbyte physical addressing space is allocated as only 1 Mbyte mapped to each NuBus slot. This allows backward compatibility with older Macintosh software. Software written for the 32-bit mode does not have this limitation.

All the Macintosh II personal computers contain a math coprocessor as a standard item (unlike most MS-DOS PCs, where it is an option). For a MC68020-based Macintosh II, the math coprocessor is Motorola's MC68881. For the MC68030-based systems, the math coprocessor is Motorola's MC68882.

As with IBM's MCA, Apple's NuBus allows multiple devices to become a bus master and contains hardware to arbitrate requests to control the bus. A device on the bus can be a potential master, capable of controlling the bus (such as an additional processor card), or a slave that cannot control the bus (such as a video adapter card). The NuBus has extra features to facilitate multiprocessing, including bus and resource locking. This can ensure a particular processor exclusive access to certain system resources (such as shared memory on a card). Figure 12-3 shows a simplified NuBus configuration with both master and slave cards. Note that the motherboard in the Macintosh II systems looks like a master card to the NuBus.

All members of the Macintosh II family of personal computers are high-performance, 32-bit systems. The original Macintosh II is based on the Motorola MC68020 CPU, with a 16-MHz processor clock. Along with all the other members of the Macintosh II family, it can support up to 8 Mbytes of RAM. It has 256 Kbytes of ROM and six NuBus slots. It includes two RS-232/RS-422 ports and a SCSI interface.

Other members of this family, starting with the Macintosh IIx, include a 3.5-inch diskette drive that is compatible with 3.5-inch drives in MS-DOS PCs. With appropriate software from Apple, data files can be exchanged with MS-DOS systems. The Macintosh IIx uses a Motorola MC68030 processor with a 16-MHz clock. The high-end Macintosh IIfx uses a MC68030 processor with a 40-MHz clock along with 32 Kbytes of high-speed static RAM, for cache.

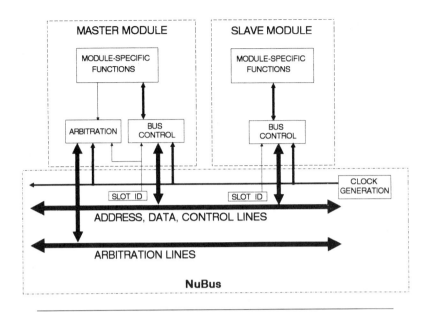

Figure 12-3 Simplified NuBus configuration for master and slave cards.

Several hardware manufacturers produce NuBus data acquisition cards for Macintosh II personal computers. Data Translation's DT2211 multifunction boards contain 12-bit analog inputs with a maximum conversion rate of 20,000 samples/second, two 12-bit analog outputs, 16 digital I/O lines, and 1 KWord of onboard memory, as a buffer for speeding up data transfers. Special software support includes Data Translation's ForeRunner driver package, for use in developing Macintosh II application programs written in C or Pascal. Data Translation's DT2221 series of multifunction NuBus cards support 16-bit (model DT2225) or 12-bit (all other models) analog inputs at conversion rates up to 750,000 samples/second (model DT2221-L, at 12-bit resolution), with two 12-bit analog outputs, two pacer clocks, 16 digital I/O lines, and 256 KWords of onboard memory. The memory buffer supports high-speed data conversion rates and enhances data transfer rates. Software support for this series includes Data Translation's TopFlight Interface Expansion Kit. This is a subroutine library, supporting data acquisition functions.

Keithley Metrabyte has several data acquisition boards for Macintosh II systems, including both NuBus and SCSI support. The MBC-625 multifunction board contains 12-bit analog inputs with a maximum conversion rate of 142,000 samples/second, two 12-bit analog outputs, three counter/timers, and 16 digital I/O lines. This board also acts as a carrier, supporting up to three optional daughter boards, for specialized functions including high-speed analog inputs (12-bit, 833,000 sample/second), high-resolution analog inputs (16-bit, 50,000 sample/second), and extra digital I/O lines. Other Keithley Metrabyte NuBus products include the low-cost, multifunction MACDIOS II JR. (with 12-bit, 40,000 samples/second analog inputs, two 12-bit analog outputs, three counter/timers, and 16 digital I/O lines) and the MACPIO-24 (with 24 digital I/O lines), shown in Figure 12-4.

Keithley Metrabyte's support software includes a driver package provided with the hardware as well as an additional function library, MBC-SFT-II. An applications software package supporting these Keithley Metrabyte cards is MBC-MI MacInstruments, which uses the data acquisition card to emulate an oscilloscope or chart recorder on the Macintosh II.

Figure 12-4 Keithley Metrabyte MACPIO-24 NuBus card. (Courtesy of Keithley Metrabyte)

National Instruments is another manufacturer supporting Apple's NuBus with data acquisition products. The NB-MIO-16 multifunction card contains 12-bit analog inputs with a maximum conversion rate of 91,000 samples/second, two 12-bit analog output channels, three counter/timers, and eight digital I/O lines. National Instrument's NB-MIO-16X uses a 16-bit ADC with a maximum conversion rate of 55,000 samples/second, along with the other features of the NB-MIO-16 card. These boards are supported by National Instrument's LabVIEW software system and their NB LabDriver package. Other National Instruments NuBus products include a low-cost multifunction card (Lab-NB, with a 12-bit, 65,000 samples/second ADC, two 12-bit DACs, three counter/timers and 24 digital I/O lines), a digital I/O board (NB-DIO-32F, with 32 digital I/O lines), and GPIB interface cards (NB-GPIB and NB-DMA-8-G).

Burr-Brown/Intelligent Instrumentation also supports the Macintosh II NuBus with a multifunction carrier board, PCI-701C (NuCarrier). This board contains 12-bit analog inputs with a maximum conversion rate of 70,000 samples/second, two counter/timers, and 16 digital I/O lines. Any of the PCI-20000 series modules can be plugged into this board, to expand or add functions (such as analog outputs).

Even though software support for Macintosh II systems still lags behind MS-DOS PCs, some data acquisition and analysis software packages are available for Apple's NuBus-based computers. One of these software products (popular on PCs) that supports multiple board manufacturers is LABTECH NOTEBOOK. As with its MS-DOS version, the Macintosh version of this program is menu-driven and easy to use. It employs menus and icons that fit into the Macintosh format. In addition, it can work with other applications, collecting data in the background.

This concludes our examination of Apple's NuBus-based Macintosh II personal computers. We will continue our overview of products for data acquisition by looking at math coprocessors next.

12.3 Math Coprocessors

In Chapter 5 we briefly touched on math coprocessors for PCs (Intel's 8087 IC for 8088 CPU machines and 80287 for 80286 CPUs). These are specialized processors that perform mathematical calculations very quickly, relieving the main system CPU of this burden. This is especially true for floating-point operations, which are very slow when performed by CPU software. Application software that uses floating-point math speeds up greatly by using such a math coprocessor, if it recognizes and supports the device. Even certain binary and BCD integer calculations are supported (and accelerated) by these math coprocessors.

Each processor in Intel's 80×86 CPU family works with its own math coprocessor. For the sake of illustration, we will only look at Intel's 8087 coprocessor for the 8088 CPU, in any great detail. The other coprocessors in this family work very similarly: the 80287 coprocessor supports the 80286 CPU, and the 80387 coprocessor works with the 80386 CPU. The coprocessor is intimately tied to the architecture of its host processor.

Figure 12-5 shows a simplified block diagram of the 8087 coprocessor. The 8087 supports four types of integer representation and three types of floating-point numerical representation (see Chapter 10 for a discussion of different numerical representation formats). The integer formats used by the 8087 are 16-bit, 32-bit, and 64-bit binary as well as 80-bit packed decimal (BCD).

The floating point formats supported are single-precision (32 bits), double-precision (64 bits), and temporary (80 bits). The 80-bit temporary format is always used internally, to minimize round-off and overflow errors, using high precision and a large dynamic range. It has approximately 19 significant decimal digits and a range of $\pm 10^{4932}$. The 8087's

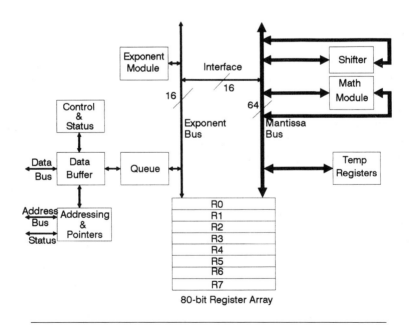

Figure 12-5 Intel 8087 math coprocessor, simplified block diagram.

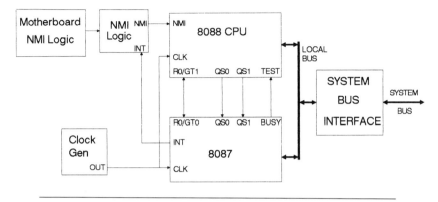

Figure 12-6 8087 math coprocessor system interconnections.

eight internal registers, used for temporary storage, are all 80 bits wide, as Figure 12-5 shows. The single-precision and double-precision formats are compatible with the IEEE floating-point standard.

The coprocessor is wired in parallel with the CPU's local address/data/control bus, as shown in Figure 12-6. Both processors use the same clock and system bus interface to other components, such as memory. The coprocessor can decode instructions simultaneously with the CPU via the queue status lines QS0 and QS1. When the coprocessor intercepts a special instruciton for itself, it asserts its BUSY line and forces the CPU to wait until it has finished executing. If the coprocessor detects an error condition, such as a command to divide by zero, it can interrupt the CPU via the NMI line (which is also used on the motherboard to indicate a memory parity error).

The 8087 supports 68 instructions, greatly increasing the command repertoire of the 8088 CPU. Besides the basic mathematical operations of addition, subtraction, multiplication, and division it can calculate trigonometric, logarithmic, and exponential functions. The coprocessor can perform floating-point calculations about 100 times faster than the CPU can (via emulation software). For example, an 8087 running at a clock frequency of 8 MHz can perform a single-precision multiplication in about 12 μsec compared to the CPU requirement of about 1 msec. The coprocessor can calculate a square root in under 25 μsec, while the CPU takes over 12 msec. Bear in mind that even a program that performs extensive floating-point math calculations will only access the coprocessor a small amount of the time, as a good deal of software overhead (such as looping control and status checking) is performed by the CPU. In a generous example, where 10% of a program's instruction are for a floating-point processor, the overall performance increase will be only 10:1 (which is 10% of the coprocessor's inherent 100:1 increase).

Luckily for most of us, we do not have to worry about the details of a math coprocessor's instructions and operating conditions to make use of the faster math processing it produces. In the simplest case, many commercial software packages will use a math coprocessor when it is present (a few packages absolutely require it to operate). If the coprocessor is not present in the PC, slower software emulation of floating-point operations is used.

Even when a programmer is developing a custom program, intimate knowledge of the math coprocessor is not required. Most high-level computer language compilers (such as C, FORTRAN, and Pascal) automatically handle floating-point operations. The programmer uses the language's standard floating-point notation and commands. At compile and link time (see Chapter 13 for a discussion of computer languages) the programmer selects the appropriate floating-point model to use. The three standard choices are coprocessor support only (8087, or equivalent required for proper operation), software emulation only (even if a coprocessor is present, it will not be used), and determine at run time (if a coprocessor is present, use it; otherwise use software emulation). The third option is a good compromise when the programmer does not know if a coprocessor will be present (or has no control over it). The drawbacks with this choice are an increase in program size (to support the "intelligent" run-time selection) and slower execution of software emulation code, compared to the software emulation-only option. However, it is usually a good choice in most cases.

The other math coprocessors in the 80×86 family are very similar in operation to the 8087, as well as retaining software compatibility. Therefore, a program written for an 8088 CPU with an 8087 coprocessor will run, unchanged, with an 80286 CPU and an 80287 coprocessor. The same is true with an 80386 CPU and 80387 coprocessor combination. The 80387 does include more features than the earlier coprocessors, including extensions to existing functions and the addition of new functions, such as calculations of sines and cosines. It also supports all aspects of the IEEE-754 floating-point standard.

There are other math coprocessors in use that are not processor-specific, as the 80×87 family is. A notable example is the Weitek 1167 coprocessor. This device is sold as part of a board-level product, for use in MS-DOS PCs, instead of an 80×87 coprocessor. Even though it is a more expensive option (since an entire board is required to use it), the Weitek Coprocessor Board (based on the 1167 coprocessor) for an 80386-based PC offers improved performance (between 7% and 68%) over using an 80387 coprocessor. Of course, you need software packages that support the Weitek chip. Some high-end mathematical software packages do support it now. There are also commercial software libraries supporting the Weitek coprocessor, for programmers.

As with Intel's 80×87 math coprocessors, Motorola's $6888\times$ coprocessors support their M68000 CPU family. These math coprocessors (68881 and 68882) are also IEEE-754 compatible. On Apple's Macintosh II personal computers, these math coprocessors are a standard item on the motherboard (as opposed to an option in the MS-DOS world). This means that all floating-point math software written for the Macintosh II can safely assume a math coprocessor is present, ensuring a consistently fast execution speed.

This concludes our discussion of math coprocessors. Next, we will take a brief look at other processor products added into a PC system. Then we will examine specialized personal computer systems, not based on the standard motherboard configurations.

12.4 Other Processor Cards

Math chips are not the only type of add-in coprocessor used in a PC, although all the other processors come as card-based products. For example, several years ago, accelerator cards were common items for PC/XT systems. These boards contained an 80286 CPU used to obtain AT performance from an older system. The system enhancement provided by these cards was limited, since the system bus was still only eight bits wide. Due to the ever decreasing costs of 80286-based PCs, these cards are not used very often anymore. There are still boards available for adding an 80386 processor to an 8088 or 80286 system, such as Intel's Inboard 386. Although similar arguments over the limited performance improvements also apply here, these boards do provide the capability of running 80386 software. However, a better alternative could be an 80386SX-based PC, which costs little more than a comparable 80286-based system and can run all 80386 software.

DSP chips are popular processors used with PCs in scientific and engineering environments. These devices are the heart of DSP boards, designed for PCs with an ISA bus. These DSP cards are special-purpose math accelerators, used to provide DSP functions at very high-speed calculation rates, minimizing the system CPU's involvement. For example, a 1024 sample floating-point FFT calculation could be performed in under 10 msec with such a card. The DSP processor is a stand-alone CPU, with its own local memory and control hardware on the card. It is not directly tied to the system CPU, as a math coprocessor is. The system CPU sends the DSP processor commands and data, via the ISA bus. The DSP processor can then operate independently of the system CPU.

The most popular DSP ICs used on these cards are the 16-bit Texas Instruments TMS320 series, whose members include the TMS32010, TMS32020, and TMS320C25. These chips are general-purpose processors, optimized for the high-speed mathematical calculations required in DSP applications. They have a special architecture with separate buses for instructions and data (called the modified Harvard architecture). This allows calculations to be preformed in parallel with instruction fetches. These devices also use *pipelining* in their computational sections, so that a current computation can progress simultaneously with a new one starting. When many consecutive calculations are done, the overall computation rate decreases dramatically.

Other popular DSP chips are supported by hardware and software products for personal computers. These include Motorola's DSP56000 series and AT&T's DSP32 series.

Most DSP hardware products come with software support, usually in the form of a library of DSP functions. A programmer can then call these functions from a conventional MS-DOS program, without worrying about the fine details of the DSP board's internal operations.

12.5   Specialized Personal Computer Systems

The standard desktop personal computer is not always well suited for harsh lab or industrial environments, due to factors such as heat, shock, dust, electrical noise, and vibration. Since more and more personal computers are finding their way onto factory floors and similar environments, many manufacturers are producing ruggedized PCs for this market.

Some of these ruggedized PCs use a standard motherboard with expansion slots, and many others use a *passive backplane*. A passive backplane is the backbone of many industrial computer bus systems (such as VME and STD BUS) as well as many mainframe computers and early microcomputers. It is simply an array of connectors, wired together to form a bus, without any active circuitry present. The CPU is on a card, plugged into the bus, just like any other expansion board (such as memory or I/O). This adds extreme flexibility to the system, since upgrading to another processor simply involves switching the CPU card. It also guarantees that all signals required by the CPU are present on the passive backplane, adding flexibility for having multiprocessors. The major penalty for this approach is added cost. However, these industrial PCs are usually much more expensive than their desktop counterparts for many other reasons.

Industrial PCs tend to use different form factors. Some systems are packed into boxes barely larger than a diskette drive, and others use large card cages for 19-inch rack mounting. The basic systems all use several expansion slots, whether with a motherboard or a passive backplane. They all have a reasonably sized power supply, a cooling fan with a dust filter, superior electromagnetic shielding (which may be poor in some desktop PCs), and often shock mounting for a disk drive (if one is present). The chassis itself may be sealed against dust and liquids.

Some industrial PCs are diskless. These systems, used only for dedicated applications, have programs stored in ROM to emulate disk-based software. This is a viable approach when a PC is embedded as part of a larger piece of equipment and a disk drive is not needed or too fragile for a harsh environment. Memory (RAM/ROM) cards that emulate disk drives are also available for conventional desktop PCs.

Most manufacturers of industrial personal computers support the PC/XT and AT buses. Most systems accept conventional PC/XT/AT cards. IBM itself has produced industrial PCs for years. As their conventional PC product line has moved into MCA, so has their industrial PC line. Currently, IBM produces both ISA and MCA industrial PCs. There are few manufacturers, other than IBM, who support the Micro Channel in industrial PCs.

A continuing problem in the area of industrial PC systems is how to enhance the ISA standard while maintaining compatibility with products from different manufacturers. A new standard, called PCXI, from Rapid Systems (Seattle, Washington) is a potential solution. PCXI is intended to be a multivendor standard for data acquisition and industrial instrumentation systems. It incorporates a standard ISA passive backplane and power supply into a modified PC chassis, which is flipped around so I/O connectors face the front of the unit. Spacing between cards is increased to 1.2 inches to accommodate metal shielding around each board. The primary reason for PCXI is to reduce the effects of PC-generated electrical noise on data acquisition and instrumentation peripherals. The backplane connections follow the ISA standard with some enhancements for better power distribution and grounding. The standard also specifies power and cooling requirements.

PCXI is the PC equivalent of VMXI, the VME bus instrumentation standard. PCXI will, we hope, be supported by several manufacturers and become established as a true standard. As of late 1990, only Rapid Systems is producing PCXI hardware.

As far as Macintosh computers are concerned, there is presently very little support for industrial systems, since Apple does not allow alternate sources for motherboards, BIOS ROMs, or systems software. The small number of industrial Macintosh-compatible systems basically repackage Apple's Macintosh II motherboards in a rugged chassis, such as 19-inch rack-mounted units. Some of these industrial Macintosh systems also include software enhancements, such as real-time operating system extensions.

This completes our overview of specialized PC hardware. In the next chapter, we will examine programming languages and the benefits of writing your own software.

CHAPTER

13

Computer Programming Languages

There may be times when you need a personal computer to perform a task not supported by commercially available software. This requires a custom program that you will have to write yourself or pay someone else to write. Unless you have very demanding requirements (very high speed operation or extremely large amounts of data to manipulate), any computer language you are familiar with should enable you to do the job. If you are new to programming, selecting a computer language can be confusing. The best approach is to choose one of the more commonly used languages (such as BASIC, C, or Pascal) that has a lot of support available for PC use. This support takes the form of a widely accepted PC version of the language, availability of many third-party software products (including function libraries and debugging programs) and a good choice of introductory books for using that language on a personal computer.

In this chapter, we will first examine some of the important distinctions between different types of computer languages as well as their similarities (especially in an MS-DOS PC environment). Then we will go on to look at a few popular languages in detail.

The only language a computer understands is its *machine language*—the binary commands telling it exactly what to do. The standard programming languages in common use convert logical constructs and instructions that make sense to people into a series of commands a computer can understand and carry out. A set of commands used to perform some desired function is considered a program.

The terms *high-level* and *low-level* are often associated with computer languages. A low-level language is very close to machine language. The most common instance of this is an *assembly language*, which does a one-to-one conversion of simple mnemonic commands into machine language commands, or *object code*. For example, in the Intel 80×86 family, there is a command to multiply two 8-bit numbers, MUL. A simple example using this command to multiply 17 and 32 (decimal) is as follows:

```
MOV  AL,17     ;load 17 into register AL
MOV  BL,32     ;load 32 into register BL
MUL  BL        ;multiply AL by BL
               ;16-bit product is in AX (AH, AL)
```

Each line of assembler code (ignoring the comments after the semicolon) is translated into several bytes representing one computer command, or opcode. Assembly language is a simpler way for a person to represent the machine language commands. The same commands in machine language (using hexadecimal notation) would be

```
B0 11
B3 20
F6 E3
```

Obviously, the assembler mnemonics make more sense than the machine language commands.

A high-level language is more abstract than a low-level language. Processor details (such as which register contains which operand) are invisible to the programmer. Only the important operations and logic are noted. A high-level language is processor-independent, making it portable. Reprogramming the above example in C produces the following commands, which will work on a 68000 CPU as well as an 80286 processor:

```
char a=17, b=32;
int  c;
c = a * b;
```

In this case, we do not need to know where the two operands are stored; we will let the compiler worry about that.

Besides high-level versus low-level, there are two other distinctions between different types of computer languages: *compiled* versus *interpreted*. In a compiled language, the program under development (an ASCII file) is translated into machine language through a separate, independent series of steps. The result of the one-time compilation process is an executable binary file (.EXE in DOS), which can then be run from DOS (or the appropriate operating system). In an interpreted language, the program is translated into machine language one line at a time, as the program is being run. Each time the program is run, it is translated into machine language again. The program has to be run from within the interpreter, which is itself a special program running under the operating system (DOS). An example of an interpreted language is BASIC. Most other computer languages, such as C, FORTRAN, and Pascal are compiled.

A compiled program executes much faster than an interpreted program, since it has already been translated into machine language. It usually requires less free memory space, since a compiled program does not need the extra overhead of an interpreter. The main advantage of an interpreted program is the flexible user interactions available. With BASIC, for example, you can control where to start, stop, and continue program execution and check variable values without modifying the program or leaving the BASIC environment. In a compiled language, these features have to be written into the program and each modification requires a new compilation process.

Of course, many debugging programs exist to assist in the development of compiled programs. A typical debugger provides an environment that allows the programmer to control execution, check variables, modify data, and perform many different tests on the program under development. An example of such a debugger, for MS-DOS systems, is Microsoft's CodeView, which supports both low-level (Assembler) and high-level (C, FORTRAN, Pascal) languages.

For compiled languages, the complete compilation process requires at least two discrete steps: compilation and linking. Under MS DOS, compilation consists of translating the original ASCII program file (the source code) into a machine language .OBJ file (the object code, or object module). The object code file is not executable under DOS. It lacks several important items of information. The linking process takes the object code and adds any library functions it requires, as well as commands that are defined in other files (or object modules).

For example, in C, the printf command displays a text string on the screen. It is a standard C library function. If an object module calls printf, this function must be extracted from the library. The linking process links the new program's object code with other object modules, from standard libraries and user-developed sources. The linker makes sure all function and variable names are defined (and do not clash) and decides where the various code modules should be located in memory. Linking adds all the remaining information DOS needs to load the finished program into memory and run it. The output of the linking process is an executable .EXE file.

The actual compilation and linking processes may each take several steps, though this is usually invisible to the program developer (such as processing the source file multiple times). Most PC linkers (including the LINK program provided with MS DOS) support many different options, such as control over where to place the completed program in memory and how to include information for debuggers. An additional step, after linking, is required to convert a .EXE program into a .COM program (as we discussed previously). If the entire program was kept within a single 64-Kbyte memory segment, it can be processed by the DOS command EXE2BIN, which converts it to a .COM file.

It should be noted here that not all computer languages neatly fit into the categories of compiled versus interpreted or high-level versus low-level. There is no doubt, for example, that C is a high-level programming language (and one of the most popular). Yet, it is a fairly "bare-bones" language having a moderately sparse set of commands, making it similar in some ways to low-level programming languages. It is the additional libraries packaged with C compilers, along with the extreme flexibility of the language, that make it so popular. C is a very efficient language, producing relatively small programs (small executable files), compared to a less efficient high-level language, like FORTRAN. It executes commands quickly, since a command in C is translated into a relatively small number of machine language commands, again making it appear similar to a low-level language.

An example of a programming language with properties of both a compiler and interpreter is FORTH. Commands (or words, as they are called in FORTH) are executed in binary (machine language) form, as in a compiled language. However, each word is executed separately under control of the environment. In addition, new words can be defined as combinations of old words. These new words are then translated into machine language before they can be executed. This type of language, having some properties of both a compiler and an interpreter, is called an incremental compiler. The commands (words) are compiled one at a time, with new ones built on combinations of existing ones.

Now we will examine a few popular programming languages in greater detail. We will start with Assembly language, the lowest level of programming languages commonly used.

13.1 Assembly Language

Assembly language (or Assembler) is a compiled, low-level computer language. It is processor-dependent, since it basically translates the Assembler's mnemonics directly into the commands a particular CPU understands, on a one-to-one basis. These Assembler mnemonics are the instruction set for that processor. In addition, an Assembler provides commands that control the assembly process, handle initializations, allow the use of variables and labels, and control output.

In the world of MS-DOS personal computers, the most popular Assembly language is Microsoft Macro Assembler, or MASM (also sold by IBM). As with most popular compilers, MASM is upgraded on a regular basis. Most of this discussion refers to version 5.0 or later, which has simplified the use of certain directives and includes support for instructions available only on 80286 and 80386 CPUs.

A *directive* is an Assembler command that does not translate into an executable instruction, but directs MASM to perform a certain task facilitating the Assembly process. An executable instruction is sometimes referred to as an op code, while an Assembler directive may be referred to as a pseudo-op code. Directives can tell MASM many different things, including which memory segment is being referred to, what the value of a variable or memory location is, and where program execution begins.

One important MASM directive is .MODEL, which determines the maximum size for a program. Remember that for an 80×86 family CPU, memory is addressed as segments, up to 64 Kbytes in length. If 16-bit addressing is used (for code or data) only a single 64K segment will be accessed. The *memory model* of a program defines how different parts of that program (code and data) access memory segments. Five memory models are supported by MASM: Small, Medium, Compact, Large, and Huge. In the Small model, all data fits within one 64K segment and all code (executable instructions) fits within another single 64K segment. In the Medium model, all data fits within one 64K segment but code can be larger than 64K (multisegment, requiring 32-bit addressing for segment: offset). In the Compact model, all code fits within one 64K segment but data may occupy more than 64K (but no single array can be larger than 64K). In the Large model, both code and data may be larger than 64K (still, no single data array can exceed 64K). Finally, in the Huge model, both code and data can be larger than 64K and data arrays can also exceed 64K.

Since larger models require larger addresses, they will produce bigger and slower programs than a smaller model will. In selecting a model for a program, try to estimate the maximum amount of data storage you'll need. Let us say you are writing an FFT program, using 16-bit integer math and a maximum sample size of 2048 points. Since each point requires two integers (real and imaginary) and each integer is two bytes long, you need 8096 bytes just to store the input (or output) data. Even if you had separate arrays for input and output data, that would still be only 16,192 bytes. As a safety margin, for temporary storage we will double this number to 32,384 bytes, which is only half of a 64K segment. It is more difficult to estimate the size of the code. In this example, we would start with the Small model. If the code turned out to be larger than 64K (which is not easy to do in Assembly language), we would move to the Medium model. These same memory models also apply to Microsoft's high-level language compilers. If you are writing a MASM program to work with another high-level language, you should use the same memory model for both.

Here is an example of a simple MASM program that displays a text string ("This is a simple MASM program") on the screen using DOS function 09h:

```
        DOSSEG                       ;Let MASM handle the segment order
        .MODEL    SMALL              ;Small model is adequate for this
        .STACK    400h               ;Set aside 1024 bytes for a stack
        .DATA                        ;Start of the data segment
text    DB        "This is a simple MASM program"
        DB        0Dh, 0Ah, 24h      ;End with CR, LF and $ char
        .CODE                        ;Start of code segment
go:     mov       ax,@DATA           ;Load data segment location into DS
        mov       ds,ax
        mov       dx,OFFSET text     ;Now DS:DX points to text
        mov       ah,09h             ;DOS string display function number
        int       21h                ;Call DOS function
        mov       ax,4C00h           ;Load DOS exit function number
        int       21h                ;Call DOS function (exit)

        END       go                 ;Start execution at label go
```

Several directives are used here. DOSSEG tells MASM to take care of the order of the various segments (code, data, stack), a detail we'd rather ignore. The directive .DATA indicates the start of the data segment while .CODE indicates the start of the code segment. The message is referred to by the label *text*, where the DB directive (Defines Bytes) indicates this is byte data (the quotation marks indicate ASCII text). The string must be terminated by ASCII character 24h ("$") for DOS function 09h. The executable instructions are placed in the code segment. The label *go* refers to the start of the program. The address of the text string is loaded into registers DS:DX. Then DOS function 09h is called to display the string. Finally, DOS function 4Ch is called to exit the program and return to DOS. The final END directive tells MASM to begin program execution at the label (address) *go*.

MASM is called a Macro Assembler, because it supports the use of macros. A *macro* is a block of program statements that is given a symbolic name that can then be used within the normal program code. A macro can also accept parameters when it is called within a program. When the source file is assembled by MASM, any macros are expanded (translated) to their original definition text. This is very handy if the same section of code, such as a programmer-defined function, is used repeatedly. Often, predefined macros may be kept in a separate file, along with other information such as variable initializations. The INCLUDE directive can read this file in, during assembly.

This brief overview of MASM has barely scratched the surface of Assembly language. Check the bibliography for other books on this subject. Again, you should write a program in Assembly language only if a high-level language is inadequate for your task. Even then, you can usually get away with just writing the most critical sections in MASM and calling them from a high-level language. Next, we will look at a popular, high-level, interpreted language: BASIC.

13.2 BASIC

BASIC is probably the most popular interpreted computer language used on PCs. This is due, in large part, to it being included with IBM-DOS and MS-DOS packages. In fact, original IBM PC systems had BASIC in ROM, to save RAM space for programs. The first IBM PC had 64 Kbytes of RAM, and a floppy disk drive was optional. If no disk drive was present, the system would start up in BASIC (since you need a disk drive to boot up DOS). PC compatible manufacturers do not put BASIC in ROM, but run Microsoft's GW-BASIC from RAM, as any other program. GW-BASIC is functionally equivalent to IBM's BASIC and BASICA.

BASIC, which is an acronym for Beginner's All-purpose Symbolic Instruction Code, was originally developed at Dartmouth College as a tool for teaching fundamental programming concepts. It is one of the easiest programming languages to learn and use. It does have serious drawbacks. Being interpreted, it executes slowly. This becomes especially obvious when it is performing a real-time task, such as controlling serial communications at high data rates. Also, BASIC does not easily lend itself to developing neat, modular programs. It is a good tool for learning, experimenting, and quickly prototyping software algorithms. It is not well suited for developing high-performance or commercial-quality software. In addition, standard BASIC can only use 64 Kbytes of memory for data and stack storage.

Interpreted BASIC has two modes of operation: *direct mode* and *indirect mode*. In the direct mode, BASIC commands and statements are executed as soon as they are entered. Results of calculations can be displayed or saved in a variable for further use. The statement or command lines themselves are lost after execution. Direct mode is useful for quick calculations or debugging operations (such as displaying or loading variable values). BASIC can accept a direct command when it is at the command level, displaying the OK prompt. An example of a direct mode command to display the result of a calculation would be

```
PRINT 23 * 17 + 2
```

The PRINT command displays the result on the screen (LPRINT sends output to the printer).

In the indirect mode, lines of program statements are stored in memory. Each program line is preceded by a line number. If a line number is missing, that command line is treated as a direct mode statement. After all program lines are entered, the program can be executed via the RUN command. The sequence of program execution starts with the lowest line number and continues through to the highest line number, unless a special statement (such as GOTO) explicitly changes the order. For example, a simple BASIC program to perform the direct mode calculation from the last example could be a single line:

```
10 PRINT 23 * 17 + 2
```

Or, it could be more generalized, with multiple lines:

```
10 A = 23
20 B = 17
30 C = 2
40 PRINT A * B + C
```

Here, using the variables A, B, and C, the numbers fed into the calculation can be quickly changed. An even better way would be to enter the variable values when the program is run:

```
10 INPUT "ENTER A: ", A
20 INPUT "ENTER B: ", B
30 INPUT "ENTER C: ", C
40 PRINT "A * B + C = "; A * B + C
```

In this case, the simple program is now general purpose. The user determines the variable values each time the program is run, using the INPUT statement, which prompts the user with the text enclosed within quotes. When the program is run, the screen would look as follows, with the operator's responses underlined:

```
RUN
ENTER A: 12
ENTER B: 7
ENTER C: 21
A * B + C = 105
OK
```

It is important to note that not all BASIC statements can be used in the direct mode (the indirect mode uses all of them), including GOSUB and RETURN for executing subroutines.

BASIC has a rich set of commands. It has a full range of mathematical and trigonometric functions, supporting both integer and floating-point calculations. It has commands for manipulating text strings, handling data file operations (supporting both ASCII and binary formats), and operator interfacing. It has several statements for program control, such as IF, THEN and FOR, NEXT. BASIC directly supports many aspects of a PC's hardware and software environment. It can read the system clock (via TIME$), directly input from or output to an I/O port (via INP and OUT), or even read from and write to system memory locations (via PEEK and POKE). BASIC provides many functions for controlling screen display, with both text and graphics (if appropriate display hardware is present). In addition, BASIC can call assembly language routines for functions it cannot directly perform (or perform quickly enough). The assembler code has to be properly written to allow interfacing to a BASIC program.

BASIC provides an environment that simplifies the process of program development. Besides entering program lines, BASIC has special commands for modifying programs. EDIT allows you to modify the specified line. RENUM automatically renumbers the program lines, which is necessary if you need to add a new program line between two existing lines with consecutive numbers. You can save a program onto a disk file (SAVE) or retrieve a previously saved program (LOAD). You can display a program on the screen (LIST) or send it to a printer (LLIST). You can even use a special trace mode (via TRON, TROFF), which displays the program line numbers as they are executed.

An important aspect of BASIC is that all the variables are global. Any part of a program can change the value of any variable. In some respects this can be handy. A subroutine does not explicitly return any value to the main program; it just writes to the appropriate variables. The flip side of this can be a problem, if you lose track of which variables are being used by which subroutines. Great care must be taken in keeping track of variables in BASIC.

Consider the following program, which averages 10 values in the array A(I):

```
10   DIM A(10)
20   FOR I = 1 TO 10
30   READ A(I)
40   NEXT I
50   DATA 1.1, 2.3, 5.7, 6.4, 2.9
60   DATA 3.0, 2.1, 4.0, 1.9, 8.4
70   GOSUB 500
80   PRINT "DATA AVERAGE = "; AVG
90   STOP
500  REM - SUBROUTINE AVERAGES I VALUES IN A(I)
510  AVG = 0
520  FOR J = 1 TO I-1
530  AVG = AVG + A(J)
540  NEXT J
550  AVG = AVG / (I - 1)
560  RETURN
```

There are several points to note in this illustrative program. Line 10 defines the data array A(I). The FOR ... NEXT loop in lines 20-40 loads 10 values into the array, from the DATA statements in lines 50 and 60. The mean value is calculated by the subroutine in lines 500-560. This subroutine is called via the GOSUB command and is terminated by the RETURN command. The values in A(I) are available to the subroutine, which first uses the variable AVG to accumulate all 10 values, with the FOR ... NEXT loop in lines 520-540. Then the average is calculated and stored in AVG, which is used by the main program in its PRINT statement (line 80).

Note that the FOR ... NEXT loops use a variable to keep track of how many times that loop is executed. Even though the range of *I* in line 20 is specified as 1 TO 10, since *I* is incremented at the end of each loop (before its value is tested), the final value of *I* is 11, when the looping is terminated. That is why the subroutine FOR ... NEXT loop, starting at line 520 loops through $I - 1$ times. If both the main program's and the subroutine's FOR ... NEXT loops used the same index variable (*I*), the program could not run properly.

BASIC has become so popular for PC use that many enhancements have been provided, making it closer to a professional-quality language. Several compiled versions of BASIC are available. You can prototype and debug a program in interpreted BASIC and then compile it, with few modifications, if any.

Some manufacturers of data acquisition products provide extensive BASIC support for their hardware. This includes assembly language driver functions that can be called from a BASIC program. Other manufacturers have produced their own enhanced version of BASIC to support their hardware and extend the language's capabilities.

As an example of this, Keithley Metrabyte sells MBC-BASIC. This is a BASIC compiler, compatible with programs written in GW-BASIC, with many enhancements. It can support the full 640 Kbytes of DOS memory, it allows for structured programming, it supports display windows, it allows the creation of reusable libraries, and it contains high-level commands for controlling some of their data acquisition hardware products.

Many other general-purpose, compiled versions of BASIC are available, such as Microsoft's QuickBasic. As the new BASIC versions evolve, they become more like other conventional, structured, compiled programming languages.

Now, we will look at a few high-level compiled languages, starting with C.

13.3 C Programming Language

C is one of the most popular general-purpose computer languages presently used by professional programmers. As we discussed previously, C combines the best features of low-level languages (ability to directly access hardware and to produce fast, efficient code) with those of high-level languages (supports abstract data structures, handles complex mathematical calculations, is well structured and maintainable).

The power and popularity of C reside, paradoxically, in its inherent simplicity. In one sense C is not very robust, because it lacks many functions present in other high-level languages, such an x^2 command. However, it contains all the building blocks to create this function along with any other high-level language operation. Many of these features are present in standard libraries that are part of a commercial C compiler package. In addition, C contains many operators not found in most high-level languages, such as bit manipulation commands. In addition, since C is modular, it is easy to add new functions as needed and use them as if they were an inherent part of the system.

C is also a well-standardized language. It was developed at AT&T Bell Laboratories during the early 1970s, by Dennis Ritchie, where it was well controlled. The language is defined by the standard text, "The C Programming Language," by Kernighan and Ritchie. Virtually all commercial C compilers adhere to the standard (although some may add enhancements, along with additional function libraries).

To illustrate some of the features of C, here is the program for calculating an average value from the previous BASIC section, rewritten in C:

```
float a[10]={1.1,2.3,5.7,6.4,2.9,   /* define data array */
             3.0,2.1,4.0,1.9,8.4};
main()                     /* Program execution starts with main */
{
float avg;                 /* variable for average value */
avg = calc_avg(a,10);      /* calculate average value of array */
printf("DATA AVERAGE = %f\n",avg);  /* display result */
}                          /* End of program */
float calc_avg(data,nval)  /* Subroutine calculates
             average value in array data, nval points long */
float *data;               /* data points to input data array */
int    nval;               /* nval contains # of values to avg */
{
int    i;      /* misc variables */
float  x;

for(i=0, x=0.0; i < nval; ++i)   /* main calculation loop */
   {
   x += *(data + i);             /* Add data values into x */
   }
x /= nval;                 /* calculate average (sum/nval) */
return(x);                 /* return result to main program */
}                          /* End of Subroutine */
```

Many aspects of C are shown in this example. Functions (including the main program and any subroutines) are specified by a name followed by parentheses, with or without arguments inside. The statements comprising the function are delimited by the braces, { }. These same braces also delimit various loops within a function or even array initialization data (for a[ ]). Program execution starts with the function main( ), the main program. When another function name appears, such as calc_avg( ), that function starts execution. When it completes, control is returned to main( ), along with a return value (if any). Statements in C are terminated by a semicolon (;), and pairs of special characters (/* */) delimit comments. Statements (and comments) can span multiple lines. C is not rigorous about text formatting in the source code. It allows programmers to format a file for easy readability. In this respect, C is a fairly free-form language. Also, it does not use line numbers, although you can give statements a label.

There are many important facets of C. One of these is *function privacy*. Any variable defined and used within a function is private to that function. Another function cannot directly access that variable. This is in sharp contrast to BASIC, where all variables are global (none are private). When a function sends a variable value to another function, it sends a copy of that variable, so the original cannot be changed by the other function. For a variable to be global, it must be defined outside of a function. In the above example, a[ ] is a global array of floating-point numbers. The only reason a[ ] was made a global array instead of a local array in main( ) was to initialize its values more easily. Also note that all variables have to be explicitly declared before they can be used. As opposed to some other languages, C must know explicitly what all the variable types are (integer, floating-point) before it can use them.

Another significant aspect of C is the use of pointers. In C, any variable (scalar or array) has two values associated with it: the lvalue and the rvalue. The lvalue is the address of a variable, and the rvalue is its actual numeric value. A pointer is used to address a variable (or a memory location). If we have a pointer, pntr1, containing the address of a variable, we can store the value of that variable in another variable, x, with the indirection operator, *, as follows:

```
x = *pntr1;
```

Similarly, if we want another pointer, pntr2, to contain the address of the variable x, we can use the address of operator, &, as follows:

```
pntr2 = &x;
```

The utility of this pointer scheme is shown in the program example above. Function calc_avg( ) defines two dummy parameters, data and nval. The parameter data is defined as a pointer to an array of floating-point values via:

```
float *data;
```

If data was just a scalar variable, it would be defined without the indirection operator:

```
float data;
```

By specifying this parameter as a pointer, we do not have to pass 10 variables to calc_avg( ). In addition, the function can handle input data arrays of variable length; it just needs to know where the array starts (via data), how long it is (via nval), and how big each element is (via the float declaration for *data).

Another aspect of pointers is that they are the only way to get around variable privacy. The only way one function can modify the rvalue of another function's local variable is if that function sends it the lvalue (pointer) of that variable. This is necessary when a relatively large amount of data must be passed between functions. Still, this is done explicitly, and the indirection operator must be used to access the variable from its pointer.

Several aspects of the notations used in C can be bewildering at first. One source of confusion is = (the assignment operator) versus == (the equality operator). The assignment operator is used to assign the rvalue of a variable, as in most high-level languages:

$$x = 10;$$

The equality operator tests a statement to see if it is true or false (in C, false is considered 0 and true is considered nonzero). So, a conditional statement, checking if x equals 10 would be

```
if (x == 10)
{
}
```

If x does equal 10, any statements within the braces would be executed.

Other notation used in the sample program may seem strange. C allows for special assignment operators, such as += or /= (as used in the sample program). The statement

    x += 10;   is equivalent to   x = x + 10;

Similarly,

    x /= nval;   is equivalent to   x = x / nval;

These assignment statements are notational conveniences. Other important operators are increment (++) and decrement (− −). As used in the sample program, the increment operator statement

    ++i;   is equivalent to   i = 1 + 1;

Similarly, − −j means j = j − 1 (decrement).

Logical operators also can be confusing. The bitwise AND operator (&) is different from the logical AND operator (&&). For example,

$$i = 0x13 \ \& \ 0x27;$$

evaluates i = 13h AND 27h as 03h (note the use of 0x for hexadecimal numbers). When used as a logical operator,

$$i = a \ \&\& \ b;$$

i is evaluated as TRUE only if both a AND b are true. The same distinctions hold true for the OR operators (| and ||).

There are two important loop control statements in C, the for loop (shown in the example program) and the while loop. As shown in the example, the for loop consists of a for( ) statement followed by one or more program statements, enclosed in braces. The for( ) statement consists of three sets of expressions, separated by semicolons: initializations (i = 0, x = 0.0), test condition (i < nval), and execute at end of loop (++i). The initializations set up a loop index variable (i) and any other variables used in the loop (x), where required. The test condition (i < nval) is evaluated at the start of each loop. This usually checks if the index is within bounds. If the test condition is true, the statements within the loop's braces are executed [x += *(data+i);]. This is followed by the executable expression (++i), usually used to increment the loop index. When the test condition is no longer true, as when the loop has been executed the requisite number of times, execution continues with the first statement following the for loop.

The while loop is simpler. It consists of a while( ) statement, which contains only a test expression, followed by braces enclosing the loop statements. If we rewrite the for loop from the example program as a while loop, we get:

```
i = 0;
x = 0.0;
while(i < nval)
    {
    x += *(data + i);
    ++i;
    }
```

The while( ) statement is a useful way to wait for an event to happen, regardless of how many times to try. If we are waiting for a device to produce data via a function get_data( ), which returns 0 if no data is present, the statement

```
while(get_data() == 0);
```

waits indefinitely until get_data( ) returns a nonzero value. Of course, in actual practice there should be a way of terminating this wait in case of error (such as a time out).

This concludes our brief overview of the C programming language, which is one of the best fits for data acquisition applications. Next, we will quickly look at a few other high-level programming languages: FORTRAN and Pascal.

13.4 FORTRAN

Many other high-level languages are commonly used for programming PCs. FORTRAN is one of the oldest, numerically oriented high-level languages, extensively used for scientific and engineering programming.

FORTRAN is an acronym for FORmula TRANslator. It is not a highly structured language (akin to BASIC), where GO TO statements are extensively used for flow control. Unlike BASIC, only lines used in branching statements get numbered.

FORTRAN supports explicit declaration of variables, but it also uses implicit variable types. It assumes that an undeclared variable is real (floating point) unless it begins with a letter between i and n (inclusive), which is an integer. For program control it uses an IF statement, GO TO statement, and the DO loop (similar to the for loop in C). To illustrate some points, here is our sample program for calculating an average value, written in FORTRAN:

```
C     CALCULATE AVERAGE VALUE
          DIMENSION D(10)
          DATA D/1.1,2.3,5.7,6.4,2.9,3.0,2.1,4.0,1.9,8.4/
          SUM=0.0
          I=1
20        SUM=SUM+D(I)
          I=I+1
          IF(I.LE.10) GO TO 20
          AVG=SUM/10.0
          PRINT,AVG
          STOP
          END
```

Note that only the line addressed by the GO TO command is numbered: 20 SUM=SUM+D(I). In the IF statement, .LE. is a logical operator (Less than or Equal). All the logical operators in FORTRAN begin and end with a period (.), such as .AND., .OR., .EQ. (EQuals), and .GT. (greater than).

The program can be simplified by using a DO loop instead of the IF( ) GO TO structure. A DO loop is similar to a for loop in C. The rewritten program is as follows:

```
C     CALCULATE AVERAGE VALUE
          DIMENSION D(10)
          DATA D/1.1,2.3,5.7,6.4,2.9,3.0,2.1,4.0,1.9,8.4/
          SUM=0.0
          DO 20 I=1,10
20        SUM=SUM+D(I)
          AVG=SUM/10.0
          PRINT,AVG
          STOP
          END
```

Now, as long as I is less than or equal to 10, any statements between the DO 20 statement and line 20 (inclusive) are executed. When this condition is no longer true, execution passes to the statement following line 20. Also note that all the variables used in these FORTRAN examples are floating point, except for the integer I.

FORTRAN is a well-established language with a large base of support. However, newer programming languages, such as Pascal and C, have superseded it in popularity, especially in the world of personal computers. It is rarely the language of choice for data acquisition or data analysis applications on personal computers, especially if low-level interfacing or graphics are involved. Still, for most applications it will work and FORTRAN is likely to continue in use for many years to come.

13.5 Pascal

Pascal is a highly structured, general-purpose, high-level language. It is another example of a computer language designed by a single person, Niklaus Wirth. It was developed as a means of teaching good programming skills and providing clear, readable, unambiguous source code. Pascal has succeeded in that goal; it is often taught as an introductory programming language to both computer science and other engineering and science students.

Pascal is a robust language. It contains all the standard mathematical operators as well as a large number of mathematical functions, such as sqrt(x), ln(x), sin(x). In addition, it contains standard procedures for data I/O and file handling. In these respects, Pascal is a higher-level language than C, which must rely on standard library functions for these capabilities.

The structure of Pascal programs is well defined. A Pascal program starts with a program declaration and is followed by declarations for constants and variables. As in C, a variable has to be declared before it can be used. If no subroutines or procedures are present, the body of the program, containing executable statements, follows. Finally, the end of the program is declared. If procedures are present, they precede the body of the main program. They are structured in a way similar to the main program. As with C, variables can be local or global. If a variable is declared in the main program it is global and accessible to any procedures defined with that program. If a variable is first declared within a procedure, it is local to that procedure (and any procedures declared within it).

Pascal has its own rules for syntax. As in C, the semicolon (;) is used to terminate statements and program sections. Comments in Pascal are enclosed within braces ({ }) and can span more than one line. The last line in a program is the end statement, followed by a period (end.). The last line of a procedure is an end statement, followed by a semicolon (end;).

To illustrate this language, we will look at our example of an averaging program, now written in Pascal. Note that in this version, input is expected from the user (via the read procedure):

```
program    Average(input,output);  {Calculates Average of
                                    10 input values}
const      Nvals = 10;             {number of values to average}
var        Sum, Avg: real;         {variables}
           Counter: integer;

procedure GetData;                 {Reads input value & accumulates}
    var  Value: real;              {local variable, for input value}
    begin                          {body of subroutine}
        read(Value);               {get data value}
        Sum := Sum + Value         {accumulate Sum}
    end;                           {end of procedure GetData}

begin                              {Start of body of main program}
    Sum := 0;                      {initialize accumulator}
    for Counter := 1 to Nvals      {accumulation loop }
        do GetData;                {call procedure}
    Avg := Sum / Nvals;            {calculate average}
    writeln('The average value = ',Avg)  {display result}
end.                               {end of program Average}
```

We see that the main program (Average) declares it uses both input (via read) and output (via writeln) functions. The standard output procedure, writeln, is equivalent to printf in C. The main program calls a procedure, GetData, which reads and accumulates the input values into the variable Sum, one at a time. GetData has one local variable, Value, used to temporarily store the input from read(Value). GetData can access Sum, because it is a global variable (defined by the main program, Average). The main data accumulation is done by the for loop, which calls procedure GetData. Also note that := is the assignment operator in Pascal. It is used to assign a value to a variable. In the constant declaration for Nvals, an ordinary = is used, since this is just defining the symbol Nvals.

Pascal is rich in control structures such as the for loop. In the example above, Counter is initialized to 1 and then incremented with each pass through the for loop until it equals Nvals. For each pass through this loop, GetData is executed. If multiple statements are to be executed within a for loop (instead of a single procedure call), a more generalized form is

```
for  Counter := StartVal to EndVal
     do begin
            {place executable statements
             here}
        end; {last statement executed in for loop}
```

This is very similar to the for loop in C, except here the index variable incrementing is implicit.

Pascal has an if ... then ... else structure, very similar to C. The argument of the *if* statement is a Boolean expression, evaluated as true or false. If it is true, the statements following *then* are executed. If not, the statements following *else* are executed, as in the following example:

```
if Value = 0
   then begin
            writeln('This is a zero value');
            {other then statements here}
        end  {last then statement}
   else begin
            writeln('This is a non-zero value');
            {other else statements here}
        end; {last else statement}
```

Pascal also has a *while loop*, functioning the same as it does in C. A Boolean expression is evaluated by the while command. As long as it is true, the statement (or loop) following the do command is executed. For example,

```
while Value >= 0
      do begin
             read(Value);
             writeln('The current value is ',Value)
             {other loop statements here}
         end; {last statement in while loop }
```

An additional control structure available in Pascal is the *repeat loop*. This can be considered the reverse of a while loop. The statement or loop following the *repeat* command is continuously executed until its exit condition, in the *until* statement, is true. Rewriting the above example with a repeat ... until structure, we get

```
repeat
    read(Value);
    writeln('The current value is ',Value)
    {other loop statements here}
until Value < 0;
```

Notice that unlike the while loop, the repeat loop is always executed at least once.

This ends our brief overview of Pascal. It is a powerful language, well suited for most programming tasks. In addition, it is well supported by compilers, libraries, and debugging tools for personal computers. We will conclude this chapter by reviewing a few key points to keep in mind when writing your own software.

13.6 Considerations for Writing Computer Programs

There are many possible approaches to writing a computer program to solve a particular data acquisition or analysis problem. Regardless of the language used, the same steps are followed in developing a usable program. A personal computer is a wonderful platform to use for software development, due to the abundance of commercially available support tools.

A necessary starting point is stating the problem and your proposed solution in general terms, written in plain English. This may be as simple as "Acquire 1024 data points, run an FFT, and report the average signal amplitude in the frequency band of 100 to 300 Hz." Next, draw a flow chart, including more of the required details (such as initializing hardware for data sampling rate and analog input range). The flow chart gives you an overview of what your program will do. It also helps you locate potential errors in logic, before they get lost in the details of the chosen programming language. Figure 13-1 contains the flow chart for a simple program acquiring 1024 data points. It flows continuously from the Start point to the End point, except for the data acquisition loop. Here, the decision box checks whether the counter has exceeded 1024. If so, it ends the program. If not, it loops back and acquires another sample.

The next step is to write the actual program and debug the source code until you can compile and link it without errors. When most compilers find a compilation error, they will point out where it is in the program (or where the compiler thinks it is), along with a clue to the type of error. An error from the linker usually means a function or a global variable was not defined. This can be due to forgetting to include a particular name in the list of object modules to link, or even a spelling error, causing a call to a nonexistent function.

Once you have an executable program, you can test it functionally with a debugging program. A source-level debugger is preferable when the

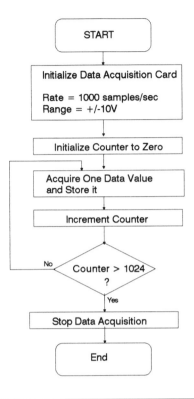

Figure 13-1 Data acquisition program flow chart.

program is written in a high-level language. This allows you to check variable values, follow the route of statement execution, and even change values to see what will happen. Most major compiler manufacturers provide a debugging program for their languages. In addition, there are third-party debugging products that support compilers from several manufacturers. As an additional aid, you can always add debugging statements to your program while writing it. These would typically display various intermediate variables or parameters returned to calling functions. You can have these statements conditionally execute, depending on the value of a global debug variable. Here is a simple example in C:

```
int   debug = 1;        /* Debut Statements enabled */
.
.
.
junk()                  /* Illustrative function */
{
int   i,j;
.
.
.
if (debug)
      printf("\nDebug values:   i=%d, j=%d\n",i,j);
}
```

If the variable debug is set to 0, the printf( ) statement in the subroutine junk( ) will not be executed.

The most critical aspect of writing software, which is commonly overlooked, is documentation. This involves adding comments to your program as you initially write it, debug it, and update it. Putting a comment on nearly every line of source code is very useful (except for extremely obvious statements, such as display outputs). It is also important to put a detailed explanation of a program at the beginning of the file. Each subroutine should also be documented at its beginning, including what it does, what its input and output parameters are, and what routines call it. Always document your programs well enough so that if you have to look at them again, several years later, you can quickly figure out exactly what you did. In the case of documentation, too much is never enough (the same point holds true for hardware designs also).

This concludes our quick survey of computer programming languages. This discussion touched on many of the major languages used on personal computers. In the next chapter we will look at some examples of data acquisition applications in the real world.

CHAPTER 14

PC-Based Data Acquisition Applications

In this final chapter, we will look at a few examples of how personal computer-based data acquisition systems are used in "real world" situations. These applications fall into three major categories, which tend to overlap: laboratory/industrial data collection, laboratory/industrial control, and embedded data acquisition and control.

In this book, we have focused primarily on data acquisition and control equipment for use in a laboratory or industrial setting. These are stand-alone systems that use a personal computer containing appropriate data acquisition cards and running software for data collection, analysis, and control. Such a system may be used for performing a laboratory experiment, obtaining automated measurements in an industrial setting, or controlling an industrial process.

Embedded applications are another way of utilizing data acquisition and control systems based on personal computers. In this case, an original equipment manufacturer (OEM) uses a personal computer-based data acquisition system as part of a larger piece of equipment it produces. The personal computer and its related hardware and software are embedded in that equipment. Usually, the software running on the personal computer is dedicated to the task the equipment was designed for. When this software is designed to insure that the personal computer always starts up in this dedicated application, it is considered a *turnkey system*. An example of this would be an automated test equipment (ATE) system, dedicated to testing particular devices (such as printed circuit boards). It is no longer

274   CHAPTER 14   PC-Based Data Acquisition Applications

usable as a general-purpose personal computer, unless the system software allows it.

We will now examine a few examples of data acquisition applications, starting with laboratory and industrial measurement systems.

14.1   Ultrasonic Measurement System

Ultrasonic waves are employed for many different types of measurements, including displacement, determination of material properties, and Doppler-shift velocity. Many of these ultrasonic applications are based on time-delay measurements. Since the speed of sound is five to six orders of magnitude slower than the speed of light (depending on the medium), the time measurements required to determine typical distances are more easily attained using ultrasonics. In air, at room temperature, ultrasonic waves travel at approximately 340 meters/second. The measured time delay t is

$$t = d/v$$

where d is distance traveled and v is the wave velocity. For a distance of 1 meter, the time delay using an ultrasonic beam would be 2.9 msec. Using a light beam, the corresponding time delay would be 3.3 nsec, which is very difficult to measure.

Ultrasonic ranging systems are commonly used to measure macroscopic distances on the order of inches to hundreds of feet. This technique is often implemented using a single ultrasonic transducer as both a transmitter and a receiver. An ultrasonic pulse is transmitted by the transducer, reflected off a target at the distance to be measured, and then detected by the same transducer. The measured time delay between the transmitted and received pulses is equal to twice the transducer–target distance divided by the ultrasonic velocity. Many low-cost ultrasonic transducers for wave propagation in air are available. One variety, an electrostatic transducer available from Polaroid Corp. (Cambridge, Massachusetts), is a popular choice for this type of application. Figure 14-1 shows a simplified implementation of this ranging system.

The sequence of events for a single measurement cycle is as follows: We start with a trigger pulse from a timing reference, which initiates a high-voltage transmit pulse that is sent to the transducer. The reflected (delayed) ultrasonic pulse is received by the transducer, goes to a receiving circuit, containing an amplifier and filter. An analog multiplexer is used to isolate the high-voltage transmit pulse from the low-voltage receive pulse (usually using diodes). An ADC starts sampling data, once the

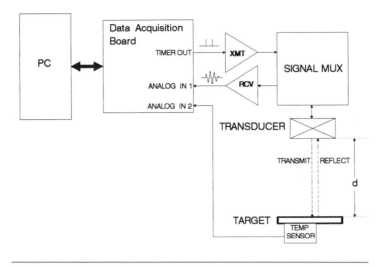

Figure 14-1 Ultrasonic ranging system.

trigger pulse occurs. The sample number containing the start of the reflection pulse multiplied by the time between samples is the time delay corresponding to the round-trip distance traveled by the ultrasonic waves.

The wavelength λ of any wave is

$$\lambda = v/f$$

where v is the wave's velocity and f is its frequency. If the ultrasonic transducer's resonant frequency is 50 kHz (as with the Polaroid transducers), the wavelength is 6.8 mm (approximately 1/4 inch). A good estimate of the displacement resolution using this technique is one-half wavelength or 3.4 mm (approximately 1/8 inch). If we needed finer resolution, we would have to use a higher-frequency transducer (such as 170 kHz for 1 mm resolution).

If we want to implement this experiment using a PC-based data acquisition system, we must first determine our measurement requirements. We will assume that a distance resolution of 1/8 inch is adequate and the maximum distance measured will be 100 feet. If we use a 50-kHz transducer, our ADC sampling rate must be at least 100,000 samples/second. To ensure reasonable data fidelity, a higher rate is preferable, such as 250,000 samples/second. The maximum distance of 100 feet corresponds to 30.48 meters. The maximum round-trip time delay is

$$\frac{30.48 \text{ m} * 2}{340 \text{ m/sec}} = 179 \text{ msec}$$

This corresponds to approximately 45,000 samples (at 250,000 samples/second). This 179-msec period also limits the maximum transmit pulse repetition rate to the inverse of that period, or 5.6 Hz in this case. This is the maximum number of transmit/receive cycles we can measure each second, without having the reflection from cycle $n - 1$ appear after the start of cycle n.

Since the data rate required for this experiment is very fast, a high-speed data acquisition card with on-board memory is called for. If a 12-bit ADC is used, 250,000 samples/second corresponds to a data transfer rate of 500,000 bytes/second, which is faster than most personal computer's DMA capabilities. This data has to be stored in the data acquisition board's local memory. For our purposes, this memory must have a capacity of at least 90,000 bytes (assuming 2 bytes/sample). In addition, this data acquisition card must have a digital output to act as the trigger line for the external transmitter as well as a data-acquisition start signal. We also want this board to have a counter/timer that can initiate an ADC conversion every 4 μsec (for 250,000 samples/second), as well as control multiple cycle timing. We also need at least a second analog input channel to periodically measure the air temperature (for velocity calibration).

There is a less expensive alternative to this relatively high-priced, high-speed data acquisition board with local memory. Since the event we wish to measure can be repetitive, instead of measuring the entire waveform in one cycle, we can acquire data over several cycles. All we need is a data acquisition board with a sample-and-hold amplifier in front of the ADC. If the amplifier has a sample window less than or equal to our sample period of 4 μsec, the conversion rate of the ADC does not matter that much. In the worst case, we acquire one sample for each waveform cycle. The overall data acquisition time will depend on the repetition rate of the transmit pulse (the overall cycle time).

To implement this, for every transmit pulse cycle, we delay the sample time of the data acquisition board by another 4 μsec. Our transmit pulse repetition rate is limited by the maximum delay time between the transmit and receive pulses of 179 msec, in this example. This means we can only generate five cycles/second. Since we need to acquire about 45,000 samples, this will take 9000 seconds, or 2.5 hours! A better way is to acquire multiple samples from each pulse cycle. Even if the ADC can acquire only 1000 samples/second, each transmit/receive cycle will produce 179 samples, now. This way, only about 50 seconds (250 cycles) is needed to acquire 45,000 samples of data.

This scheme is shown with the waveform in Figure 14-2. It would acquire samples spaced 1 msec apart, from a single transmit/receive cycle, as represented by the "X" symbols. During the next cycle, the acqui-

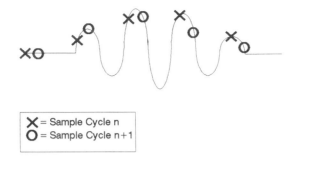

Figure 14-2 Using multiple cycles to acquire a repetitive waveform.

sition starts 4 μsec later, as shown by the "O" symbols. This process continues until the entire waveform is filled in. Since the window of the sample-and-hold amplifier is no more than 4 μsec, it is equivalent to acquiring data at 250,000 samples/second, except it takes more time to acquire all the data, and it is done over several transmit/receive cycles. The separate data acquisition cycles must be interleaved by the computer to produce the completed waveform.

If the final desired data will reside in an array D[45,000], using the multiple cycle scheme, we will assume the first transmit/receive cycle (cycle 0) has no offset time, the next cycle (cycle 1) starts 4 μsec after the trigger pulse, and so on, until cycle 249 starts 996 μsec after the trigger pulse. Note that the data at time = 1000 μsec (1 msec) is the second point from cycle 1, since these data are all 1 msec apart. If each of these cycles produces a data array Cn[180], the reconstructed waveform data D[m] (where m = 0 to 44999) will be

D[0] = C0[0], D[1] = C1[0], . . . ,   D[249] = C249[0]
D[250] = C0[1], D[251] = C1[1], . . . ,   D[499] = C249[1]
.   .
.   .
.   .
D[44750] = C0[179], D[44751] = C1[179], . . . , D[44999] = C249[179]

Once the waveform data array is acquired and reconstructed, it can be analyzed. Figure 14-3 shows a typical waveform from an ultrasonic ranging system. Since the transducer is multiplexed for transmit and receive signals, ringing from the transmit pulse appears in the acquired waveform. Since this transmit signal is fairly constant, we can ignore the data for the first millisecond or so. This *lockout window*, corresponding to

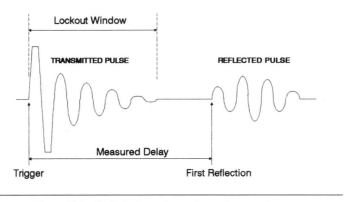

Figure 14-3 Typical ultrasonic ranging system waveforms.

about 1/2 foot, limits the minimum distance that can be measured. Any reflected pulse arriving within this window will be obscured by the transmit signal. If we used separate transmit and receive transducers this would not be a problem.

The lockout window can be implemented either in the analysis software or by initially starting data acquisition after the nominal 1-msec window period (and adding that time offset to the collected data). Since only about 250 data samples would be saved this way, the software approach is better, since it allows for adjustment of this window after data has been acquired.

One important point to keep in mind when attempting accurate ultrasonic measurements is that the velocity of ultrasonic waves is a function of temperature. That is why the ranging system in Figure 14-1 uses an additional analog input channel to measure the air temperature. This temperature measurement does not have to be done very often—once per acquired waveform is more than enough. The relationship between the speed of sound in air v and the temperature of the air T (in degrees kelvin) is

$$v = 331.4 * \sqrt{(T/273)} \text{ m/sec}$$

Other environmental factors, such as relative humidity and barometric pressure, have a much smaller effect on ultrasonic velocity and can usually be ignored. Relative humidity does have a large influence on the attenuation of ultrasonic waves.

Depending on how the acquired data looks, analysis can be fairly simple or very involved. If the reflected pulse's signal-to-noise ratio is high and its first peak is readily observable, the analysis simply consists of finding the location of that peak, which corresponds to the round-trip time delay of the ultrasonic pulse. This could be a peak-detector algorithm, checking data values with an amplitude greater than a specified noise threshold.

Of course, in the real world, things are rarely this easy. One complication is the attenuation of the ultrasonic waves. As the target distance increases, the amplitude of the reflected pulse decreases (as does its signal-to-noise ratio). It becomes more difficult to discern the first peak of the reflected pulse. An added complication would be an imperfect, rough-target surface causing scattering of the ultrasonic pulse, resulting in a "fuzzy" echo. This is because different (spatial) portions of the reflected ultrasonic pulse arrive back at the transducer at slightly different times, causing the resulting echo to be spread out in the acquired waveform.

Various DSP techniques can be used to solve these problems. Implementing digital filtering in software can help eliminate noise and enhance the reflected pulse. Another approach is to calculate the FFT of the waveform and measure the slope of the resulting phase curve, in the frequency domain. This phase slope is proportional to the absolute delay of the reflected pulse, with a time offset of half its width. This offset can be determined by shifting the original waveform so that the reflected pulse starts at time = 0 and then calculating its FFT. Subtract this phase slope from the phase slope of the unshifted waveform's FFT, producing the corrected time-delay phase slope. This analysis can be done with many different commercial software packages, requiring little or no programming.

This same experimental setup can also be used to directly measure the thickness of a material sample, with a resolution determined by the frequency of the ultrasonic transducer, as long as the speed of sound through that material is known. Whenever an ultrasonic beam passes through an interface between different mediums, such as air and a solid, there is a change in acoustic impedance and some of the beam is reflected at the interface.

As shown in Figure 14-4a, if a transmitted ultrasonic pulse hits a material of thickness d, some energy is reflected from its front surface (if its acoustic impedance differs from the surrounding media), resulting in the first echo. The rest of the beam passes into the material. Some of that beam is reflected from the back surface (the rest continues out the back). Part of the beam reflected from the back surface now passes through the front surface, back to the transducer, resulting in the second echo. The remainder of the beam reflects back into the material again, eventually resulting in the third echo. This process continues with multiple reflections.

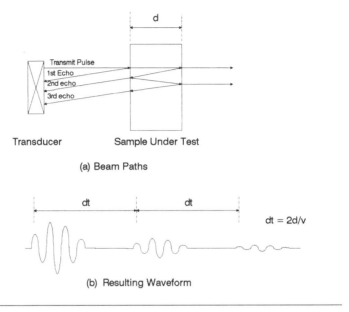

Figure 14-4 Using multiple reflections for thickness measurements.

Each successive echo is lower in amplitude, since energy is lost at each reflection (even if the material had negligible attenuation). The time delay between successive echoes, as shown in Figure 14-4b, is

$$dt = 2d/v$$

since each echo is separated in time by a round trip through the material thickness. The measurement of dt can be done fairly accurately using autocorrelation. Since each echo pulse is basically an attenuated version of the previous pulse, calculating the autocorrelation of this waveform (the cross correlation with itself) will produce peaks at

$$t = 0, dt, 2\ dt, \ldots, n\ dt$$

depending on how many echoes appear in the waveform. Each successive autocorrelation peak will be lower in amplitude than the previous one; however, the peak located at time dt is a good measure of $2d/v$.

If the correlation peaks are too broad to give an accurate measurement of dt, that curve can be differentiated. The zero crossing point of the first derivative of the autocorrelation curve is an accurate location for the peak of dt.

We should note that if the sample thickness is known, the velocity of sound through the material can be calculated from this same multiple reflection measurement. In addition, various material parameters (such as elastic modulus) can be determined from this data.

This concludes our example of an ultrasonic ranging system. Next, we will turn our attention to another example, implementing an electrocardiogram (ECG) measurement system using PC-based data acquisition products.

14.2 Electrocardiogram (ECG) Measurement System

The acquisition and analysis of human electrocardiograms (ECGs) is of great interest to many medical researchers. The ECG is a graph of voltage variations produced by the heart muscle and plotted against time. Automated analysis of ECG data is an active area of ongoing research, as a means of improving diagnosis and prediction of heart disease. The requirements for implementing an ECG data acquisition system using a personal computer platform are very different from the previous example of an ultrasonic ranging system.

ECG data consists of very low-frequency components. Most of the spectral content of an ECG fits within a bandwidth of around 10 Hz, with very little energy present above 100 Hz. A typical diagnostic ECG recorder has a bandwidth of 0.05–100 Hz. Hence, very low data-acquisition rates are used—typically 250 samples/second, to ensure good fidelity. The amplitude of ECG voltages are very low, in the range of tens of microvolts up to several millivolts. The transducers used to detect these voltages are electrodes placed on the surface of arms, legs, and chest. They connect to isolation circuitry, to protect the patient from any current that may be produced by the ECG recording equipment. Then the ECG signals must pass through differential amplifiers, required to provide high gain and good common-mode noise rejection. If ECG data will be digitized, it is commonly connected to an anti-aliasing filter, with a 100-Hz bandwidth.

Even though the acquisition rates for ECG data are relatively low, the volume of data recorded for research purposes tends to be extremely large. It is common for medical research projects to acquire several hours of ECG data from each subject, sometimes for as long as 24 hours. This data usually consists of two channels of 12-bit or 16-bit readings at a conversion rate of 250 samples/second. If we assume that no data compression is applied, we require 4 bytes of storage for each sample interval (for two channels) for a data storage rate of 1000 bytes/second. If we record 1 hour of data from a patient, it will occupy 3,600,000 bytes. Acquiring data from many subjects or recording several hours from each one will obviously use up a large amount of memory storage very quickly. It is no wonder that data compression techniques are routinely applied to ECG storage problems.

Figure 14-5 shows one beat of a typical, normal ECG waveform. Various components of an ECG cycle (one beat) have specific names. A beat starts with the P wave, which represents the original electrical impulse in the heart, beginning the cycle. It usually has a small amplitude. The QRS complex, consisting of the Q, R, and S waves, is usually the largest amplitude component in an ECG. The Q wave, itself, may have a very small amplitude (sometimes it is unmeasurable), while the R and S waves can be quite large. The cycle ends with the T wave, representing the electrical recovery phase of the heart, preparing it for the next beat.

Clinically significant information is obtained from an ECG by measuring several parameters, such as the relative amplitude, width, and time duration of these component waves, as well as the time between the components. In addition, the time between beats is important as a measure of instantaneous heart rate. Occasional, abnormal beats are also looked for, as indicators of potential problems. This requires a means of categorizing the data on a beat-by-beat basis, as either normal or abnormal.

Figure 14-6 shows a simple block diagram of a PC-based ECG recording system. The data acquisition board only needs to provide a throughput of 500 samples/second, assuming two channels of data digitized at a rate of 250 samples/second. A 12-bit ADC will provide adequate resolution. For this application, the analog front end is very critical. Elec-

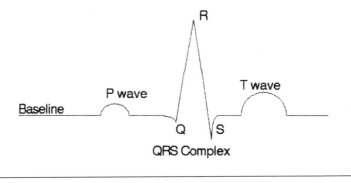

Figure 14-5 Typical normal ECG beat cycle.

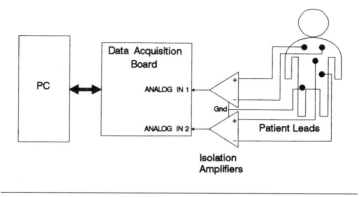

Figure 14-6 PC-based ECG recording system.

trical isolation must be provided between the patient electrodes and the data acquisition system. Any ground current flowing from the measurement system to the patient could be a serious health hazard, causing defibrillation (from a shock directly to the heart). Therefore, isolation amplifiers are used. These amplifiers need to have differential inputs.

The amount of gain provided by the isolation amplifiers will determine the analog input range required for the data acquisition card. If the isolation amps serve strictly as buffers, then high-gain, differential analog inputs would be needed. These isolation amps should also have differential outputs, so any common-mode noise on the wires connecting them to the data acquisition board's analog inputs will be rejected. Of course, the data acquisition board's analog inputs must be differential. In this case, a 12-bit ADC board with a nominal input range of ± 5 V and variable gain, up to $500\times$, would be a good choice. At the maximum gain ($500\times$), the analog input range is ± 10 mV, with a resolution of approximately 5 μV. Most ECG waveforms will fit within this range.

A better arrangement would be to provide most of the analog gain with the isolation amplifier, in a separate electronic module. This would minimize the effects of noise pickup in the cable connected to the data acquisition board, as well as noise within the PC itself. In general, it is always a good idea to implement high analog gain outside of a PC, whenever possible. This would allow the use of a simpler, low-gain analog input board. Differential inputs would still be preferable, but no longer mandatory.

In either case, a relatively slow ADC is adequate, at moderate (12-bit) resolution. DMA capabilities are not required, since the maximum data-transfer rate would be only 1000 bytes/second (since the overall acquisition rate is 500 samples/second). The data acquisition card should have a counter/timer to produce the data conversion clock. An analog output (DAC) would only be necessary if stored and analyzed data will have to be produced in analog form, at some later time. An example would be to simulate a real-time ECG signal for testing another piece of diagnostic equipment.

Most commercial data acquisition software packages would not have any problem storing data acquired at these low rates. The personal computer used must have a large hard-disk drive to accommodate the big data files produced by the relatively long experimental runs. As we saw previously, one hour of data requires approximately 3.6 Mbytes of storage. Even an 80-Mbyte drive would fill up quickly with this amount of data. It could not hold 24 hours of data (about 86 Mbyte).

One choice here is to use a tape drive, either to back up data from a large hard drive or to directly store data as it is acquired. Special software is required to deal with a tape drive, and most data analysis packages will only work with data on disk files. Therefore, using a tape drive to back up conventional-disk data files is a simpler approach. Otherwise, you may have to write a lot of your own software for storing and retrieving data on tape.

Another way of dealing with this problem of how to store such large amounts of information is to use data compression. Much research has been done on using different data compression techniques on ECG data. Unprocessed ECGs contain a large amount of redundant information. A large fraction of the data is simply the constant baseline between consecutive beats. Linear predictors can provide reasonably large compression ratios, without excessive distortion, if data acceptance windows are carefully selected.

Another aspect of the redundant nature of ECG data is that most beats from the same patient look very similar, often nearly identical. Only the occasional, abnormal beat appears significantly different. One way of exploiting this characteristic is to apply statistical methods to the data compression problem. Since most ECG data tends to have a lot of straight lines and smooth amplitude variations, it is well suited to delta encoding. Only the amplitude difference between adjacent points is stored, as a small number. If these delta values are calculated from a representative data sample, for a particular subject, they can be statistically analyzed. Then Huffman codes, only a few bits long, could be applied to the most probable delta values.

If most of the original 12-bit data can be represented as 4-bit Huffman codes for delta values, the overall compression ratio would be about

3 : 1. Most data from one patient is likely to follow the same distribution of delta values and have a similar compression ratio. As we saw previously, this is Delta Huffman Encoding. Even though it does not produce very high compression ratios, the restored data is completely identical to the original waveform, with zero distortion.

An enhancement to Delta Huffman Encoding is to identify all baseline data points that fall within a window of constant amplitude (such as noise variations). These points can be replaced by their average amplitude and the length of this line, using a special escape code in the delta Huffman data stream. This addition could increase the compression ratio by another factor of 2 (to around 6 : 1) while resulting in a small loss of fidelity. Depending on the window size used, an RMS distortion of less than 1% is easily attainable at these compression levels.

This enhancement is effectively implementing a zero-order predictor (ZOP), along with Delta Huffman Encoding. The algorithm used should carefully decide when to use the ZOP instead of Delta Huffman codes, for maximum bit savings. Since the delta values we would replace in this case would be zero, or close to it, they would probably require only two or three bits in their Huffman code. For now, we will assume they use three bits per point. If the escape code is 8 bits, the average amplitude is 12 bits and the line length is 8 bits (allowing for a line representation over 1 second long, at 250 samples/second), it will take 28 bits to represent this straight line. Therefore, this approach saves storage space if the line is more than nine points (36 msec) long, corresponding to $9 * 3 = 27$ bits.

One of these data compression methods could be applied to previously acquired data, already stored in disk files. Several commercial software packages are available that provide data compression for all types of files found on a PC. Since these products are designed to work with any file type, they produce no data distortion (since most files cannot tolerate any change in their contents, such as a program or a document). As a result, they produce fairly low compression ratios, typically 2 : 1.

A better approach is to use software specifically designed to compress ECG files or to implement data compression in real time, as the data is being acquired. Since the data transfer rate for this example is relatively low, it is possible to retrieve readings from the data acquisition card as a background task, using hardware interrupts. This would allow the PC to use its spare processing time on the foreground task of data compression. Raw (unprocessed) data would be stored temporarily in RAM, until it is compressed and written to a disk file.

Besides data compression, other analysis techniques are applied to digitized ECG data, usually for diagnostic purposes. This analysis usually involves measuring amplitude, time, and shape parameters of various portions of each beat, to place it in a diagnostic category. For example, a beat that occurred much earlier than expected, based on several previous beats, would be classified as premature and might have medical significance. Other analyses may employ FFTs or other transforms, as well as correlation techniques.

Since this analysis is very specific to ECG data, some programming would probably be required. An entire program could be written in a general-purpose language, such as Pascal or C. Alternatively, a commercial data analysis package could be employed to test out various algorithms. Even here, some programming may be required to implement the desired algorithm, such as with ASYST, or a similar product.

This concludes our look at using a personal computer for implementing a system for acquiring and analyzing ECG data. Next, we will look at an example of employing an embedded PC in a commercial product using data acquisition and control functions.

14.3 Commercial Equipment Using Embedded PCs

So far, we have considered stand-alone personal computer systems, configured for data acquisition tasks. Many equipment manufacturers require data acquisition and analysis functions in their end product. One approach, becoming increasingly popular, is to use an *embedded PC* as a major component of the product. Depending on the manufacturer's requirements, the personal computer may still function as a general purpose PC and run most commercial software packages. On the other hand, it can be completely dedicated to the tasks required by the overall product it is part of and be unable to run any general purpose software. In this case, the embedded PC may not even have a floppy disk drive. It could run programs from ROM storage, configured to look like a disk drive to DOS.

There are many advantages to using an embedded PC in a commercial product, especially for data acquisition, analysis, and storage functions. There is a huge amount of commercial hardware and software products available, minimizing in-house development costs as well as a new product's time-to-market. Compared to other industrial computer hardware (such as VME and STD BUS products), personal computers and their support products are less expensive, more readily available, and offer more "user friendly" development tools.

This trend toward using embedded PCs is reflected by the increase in the number of products for this market. Many manufacturers now produce miniaturized PCs, based on a single board, designed to fit within a product using a minimum volume. These products can support standard

PC peripherals, such as disk drives and displays, yet will work without them for a scaled down version in the final product. Software development for products using these embedded PCs can be carried out on standard desktop personal computers.

If an embedded PC is technically adequate for performing the required task and its cost can be justified, it is often a good choice, especially compared to other dedicated computer systems. If a PC would be grossly under-utilized in an application or the product is very cost sensitive, a dedicated CPU or a microcontroller board is a better alternative.

14.3.1 The CYBEX 340 Extremity Testing System

As an example of a typical use of an embedded personal computer in a piece of commercial equipment, we will look at the CYBEX 340 Extremity Testing System. CYBEX, a division of Lumex Inc., manufactures

Figure 14-7 CYBEX 340 Extremity Testing System. (Courtesy of CYBEX, a Division of Lumex, Inc.)

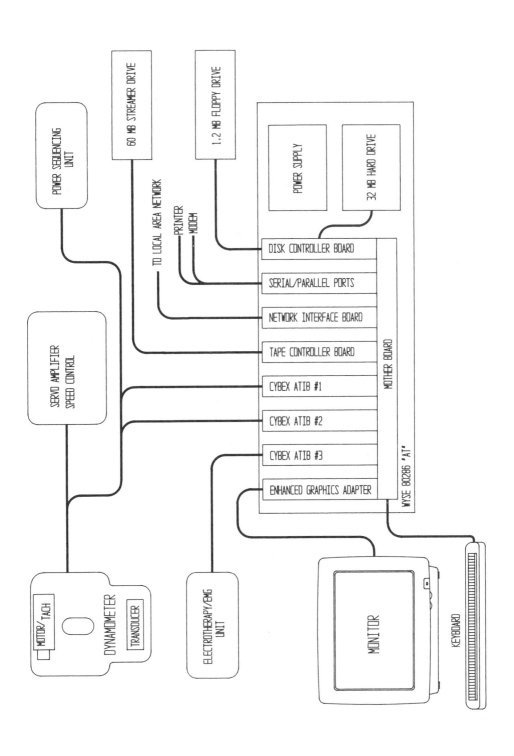

testing and rehabilitation equipment for the fields of sports medicine, physical therapy, and fitness. These machines are used for evaluating and improving human athletic performance and fitness as well as aiding injury recovery. The 340 System is used for the testing, exercise, and rehabilitation of the extremities (arms and legs).

The CYBEX 340 system is a large piece of equipment, which incorporates a full-sized desktop PC chassis in its electronics cabinet, as shown in Figure 14-7. The personal computer used is a Wyse 286 PC, which is an AT (ISA) system, based on an 80286 CPU with a 10-MHz clock and containing 640 Kbytes of RAM. The keyboard, monitor (EGA), floppy drive, and streaming tape drive are mounted externally, for normal user access. The PC also contains a 32 Mbyte hard disk drive and runs MS DOS. Other standard PC peripherals used in this system are a parallel port, connected to an external printer, and a serial port, connected to an internal modem. It contains a 60-Mbyte tape drive unit, for backing up the large amounts of data collected and stored in its database system. In addition, an optional network interface card may be present, to connect the system to a Local Area Network (LAN) of other CYBEX systems and PCs, including a PC set up as a *file server*, controlling the network.

By using standard peripherals, retaining a floppy drive, and running software under MS DOS, the PC embedded in the CYBEX 340 system can function as a stand-alone personal computer and run most commercial software packages. In addition, it runs custom CYBEX software, used for motion control of the extremity testing equipment, acquiring and storing patient data, producing reports, and other functions aiding the medical practitioner.

Figure 14-8 shows a simplified block diagram of the CYBEX 340 system. It is clear from this diagram that, at least electronically, the embedded PC is the heart of the system. The mechanical heart of this CYBEX system is its dynamometer. This unit contains a motor, controlled by a switching servo amplifier, which drives a series of clutches coupled to an output shaft connected to the patient's limb.

The speed of the motor limits the maximum speed the patient can move his or her limb. A patient trying to move faster than the set speed would produce more torque but maintain that fixed speed. The measured torque, at this constant speed, produces clinically significant information about the health and strength of that limb. For example, if someone had a

Figure 14-8 Block diagram of the CYBEX 340 Extremity Testing System. (Courtesy of CYBEX, a Division of Lumex, Inc.)

knee injury that was manifested at a particular point in that joint's range of motion (the maximum range that joint can rotate, measured in degrees), there would probably be a drop in torque output at that location. The data collected by this system is torque versus angular position. The torque is a measure of the force produced by the muscles of the tested limb. The angular position covers that limb or joint's range of motion. A typical example of data produced by a CYBEX 340 system (measurements of the author's knee) is shown in Figure 14-9.

For technical and economic reasons, CYBEX chose to produce its own data acquisition boards for the 340 system, instead of adapting general-purpose commercial cards. These boards plug into the standard ISA bus of the Wyse PC, as any commercial card does. As shown in Figure 14-8, these custom boards connect to specialized system hardware, such as the dynamometer, servo amplifier, and power sequencing unit.

The functions available on these custom boards include analog input, analog output, digital I/O, and counter/timers. The analog input is the torque signal from the dynamometer. The torque signal is derived from a capacitive load cell (a pressure transducer) hydraulically coupled to the dynamometer shaft. This signal, having a 10-volt dynamic range, is digitized by a 10-bit ADC. The overall torque range is 360 foot-pounds, so the system can resolve torque as low as 0.35 foot-pounds (or approximately 4 inch-pounds).

The ADC used in the CYBEX 340 system has several modes of operation. Conversions can be triggered asynchronously or at a fixed rate, under software control (up to 25,000 samples/second), as in a conventional, time-based data acquisition system. However, the data of interest to this system is torque versus angular position. To directly measure this, the ADC conversions are normally triggered by an optical encoder, coupled to the dynamometer shaft. This encoder acts as an angular position transducer, producing a pulse for every 1/2 degree of dynamometer shaft rotation. This produces torque versus position data, independent of rotational velocity, eliminating unnecessary data conversions at slow speeds. This feature could be implemented with a general-purpose, commercial data-acquisition card if it accepted external trigger signals.

An analog output of the CYBEX board is used to control the motor speed. This speed control voltage (SCV) is a 0–10-volt signal, sent to the servo amplifier, which drives the motor. The motor contains a tachometer that produces an analog voltage proportional to its speed. This tachometer signal is sent back to the servo amp, to complete the feedback loop required for precise speed control. The motor speed set by the servo amp is proportional to the controlling SCV.

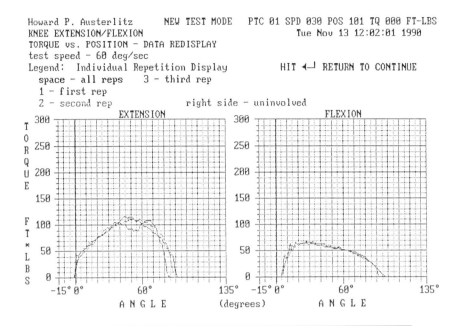

Figure 14-9 Typical data display from a CYBEX 340 system. (Courtesy of CYBEX, a Division of Lumex, Inc.)

The SCV is derived from a 10-bit multiplying DAC. Ideally, since the motor speed range is 0–500 degrees/second, we would like a linear conversion of a digital speed word (going into the DAC) to the actual motor speed. This would mean a digital speed setting of 0 would produce 0 degrees/second, a setting of 1FFh would produce a speed of 250 degrees/second, and a setting of 3FFh would produce a speed of 500 degrees/second. This transfer function of output motor speed versus input digital word should be a straight line passing through the origin.

Due to hardware variations in the motor, servo amplifier, and DAC, the actual transfer function may have the wrong slope or intercept (offset). To correct for this, the system is self-calibrating. It can measure the motor velocity via an *optical interrupter* (a simple optical encoder). The 10-bit multiplying DAC receives an analog input from an 8-bit calibration DAC. This 8-bit DAC is only written to during the calibration procedure and is used to adjust the slope of the transfer function. The result is highly accurate motor-speed control. If a commercial data acquisition card was used to implement this calibration feature, it would require two DACs, at least one being a multiplying DAC, or one very high-resolution DAC.

The counter/timers available on the CYBEX boards are based on an Intel 8254 IC. They are used for various system timing and counting functions, such as measuring motor speed during calibration. Many different functions are implemented using digital I/O lines. One example is using software-controlled gating of several possible sources into one hardware interrupt line. Another is the remote power-on capability of the system.

The CYBEX 340 system has a sophisticated power supply controller (the Power Sequencing Unit in Figure 14-8). Part of the system, including the modem, is always powered on. This is ensured by using an Uninterruptible Power Source (UPS), to protect against AC power-line problems. Under normal operating conditions, the rest of the system, including the PC, is powered on via a front panel switch. However, if the modem receives a valid carrier signal (over the telephone lines), it will automatically power up the system via its Carrier Detect signal. When the PC boots up, part of the system software that is always run (via a call from the AUTOEXEC.BAT file) checks to see if the system turned on due to normal switch operation or from modem activation. If it was due to the modem, special software enabling remote diagnostics, data transfers, or software updates can be run. The use of a standard PC in the CYBEX 340 system makes the implementation of this "intelligent" remote turn-on feature fairly straightforward.

The advantages and features of an embedded personal computer, illustrated by the CYBEX 340 system, apply to a wide variety of computer-controlled equipment, even if a product must be fairly small; the size of PCs designed for embedded applications is continuously shrinking. We are likely to see increasing growth in commercial equipment utilizing embedded PCs. An additional advantage of using an embedded PC is that the final system can be functionally prototyped using commercially available data-acquisition hardware and software products, even if it will eventually use custom boards and programs. This can certainly help shorten the development cycle of a new product. At the very least, it provides a software development group with hardware to work with while the final data acquisition boards are still in the design process.

14.4 Future Trends in PC-Based Data Acquisition

This concludes our examination of a few "real-world" examples of data acquisition systems based on personal computer platforms. This field is constantly changing, with new products, standards, and approaches appearing continuously.

In the realm of IEEE-488 (GPIB) equipment, for example, evolutionary changes continue. The original IEEE-488 standard, released in 1975 (now referred to as IEEE-488.1) specified the hardware interconnections between GPIB equipment, but did not address many areas of protocol, such as data formats, status reporting, common device commands, or error handling. These features varied from one manufacturer to the next, complicating the software development process for GPIB systems. In 1987, the IEEE-488.2 standard defined these protocols. However, this standard still does not address the problem of every instrument having a unique set of commands for its specific functions. Recently, a consortium of major GPIB equipment manufacturers (including Hewlett-Packard, Tektronix, National Instruments, and Keithley Instruments) developed a superset of IEEE-488.2, called Standard Commands for Programmable Instrumentation (SCPI). SCPI defines a comprehensive command set that can be used by all automated instruments. All standard instrument functions are included in this command set. For example, the command to measure a voltage is :MEAS:VOLT?. This same command will work with any automated instrument capable of producing a voltage measurement. If SCPI becomes an accepted standard, GPIB programs could finally become truly portable and universal.

The power and cost of personal computers is another area that will undoubtedly show continued improvements. In the MS-DOS PC world, faster and more functional computers based on 80386 and 80486 CPUs (and their successors) will continue to appear, along with hardware and software products that exploit their features. The huge lag between the appearance of a new CPU (along with computer systems incorporating it) and software that makes use of its advanced features is likely to continue or even grow. This should help prevent the most advanced PC platforms from becoming obsolete too quickly. Only in recent years have 8088-based PC/XT systems been considered obsolete for many applications (yet they are still being produced by some manufacturers, albeit at much lower volumes than previously). Today's 80386 and 80486-based PC is unlikely to be considered obsolete for a long time to come.

IBM's Micro Channel will undoubtedly continue gaining support (with PS/2 and compatible PCs) from hardware manufacturers. For now, Micro Channel systems are coexisting with ISA (AT) systems. The ISA bus is not likely to become obsolete in the near future, regardless of the eventual fate of the EISA standard.

In the world of Apple's Macintosh family of personal computers, the changes started in recent years are likely to continue. The "open architecture" of NuBus, supporting independent manufacturer's boards for a variety of fields, including data acquisition, should continue. The remaining problem with the Macintosh line is the absence of compatible systems ("clones") from other computer manufacturers, to provide competition over price and features.

In the field of sensors, a growing trend is toward integrating more functions in a sensor unit. This usually applies to signal conditioning, where a sensor output is suitable to directly connect it to an ADC. Sensors with some local "intelligence" may also become more common. The increased use of electronic sensors in major consumer products such as automobiles will continue to advance this field rapidly.

It is nearly impossible to accurately predict future trends in the personal computer and data acquisition industry. If you are putting together a PC-based data acquisition system, stick to your current requirements, with an eye on future needs. You should be aware of trends in the industry, but relying on new, untested technologies (or companies, for that matter) can be a big gamble. Always try to first use current, established products to solve a problem. It will usually cost you less time, money, and frustration.

APPENDIX **A**

Data Acquisition Hardware Manufacturers

This appendix contains listings for manufacturers of data acquisition hardware usable on personal computers. Hardware manufacturers who also supply software products are listed here, rather than in Appendix B, since their software is usually tied to their hardware products. Under the listing of PLATFORMS SUPPORTED, PC/XT/AT refers to both IBM and compatible systems. PS/2 refers to Micro Channel systems.

ACCESS
9400 Activity Road
San Diego, CA 92126
(619) 693-9005

PRODUCT LINE Analog I/O, Digital I/O, RS-422/485 Interface
PLATFORMS SUPPORTED PC/XT/AT

ADAC Corp.
70 Tower Office Park
Woburn, MA 01801
(617) 935-6668

PRODUCT LINE Analog I/O, Digital I/O, Software
PLATFORMS SUPPORTED PC/XT/AT

Analog Devices
One Technology Way
P.O. Box 9106
Norwood, MA 02062
(800) 426-2564

PRODUCT LINE Analog I/O, Digital I/O, Signal Conditioning, Software
PLATFORMS SUPPORTED PC/XT/AT, PS/2, other industrial computers

Analogic Corp.
360 Audubon Rd.
Wakefield, MA 01880
(800) 446-8936

PRODUCT LINE Analog I/O, Software
PLATFORMS SUPPORTED PC/XT/AT

Burr-Brown/Intelligent
Instrumentation
1141 W. Grand Road
M/S 131
Tucson, AZ 85705
(602) 746-1111

PRODUCT LINE Analog I/O, Digital I/O, Signal Conditioning, GPIB Interface, Software

PLATFORMS SUPPORTED PC/XT/AT/EISA, PS/2, Macintosh II

Capital Equipment Corp.
99 S. Bedford St.
#107
Burlington, MA 01803
(617) 273-1818

PRODUCT LINE GPIB and RS-232 Interface, Instrument Control Software

PLATFORMS SUPPORTED PC/XT/AT

Computer Boards, Inc.
44 Woods Ave.
Mansfield, MA 02048
(508) 261-1123

PRODUCT LINE Analog I/O, Digital I/O, GPIB Interface

PLATFORMS SUPPORTED PC/XT/AT

Contec Microelectronics
USA, Inc.
2010 N. First Street
Suite 530
San Jose, CA 95131
(408) 436-0340

PRODUCT LINE Analog I/O, Digital I/O, GPIB and RS-232/422 Interface, Signal Conditioning, Software

PLATFORMS SUPPORTED PC/XT/AT, PS/2

Data Translation, Inc.
100 Locke Drive
Marlboro, MA 01752
(508) 481-3700

PRODUCT LINE Analog I/O, Digital I/O, Signal Conditioning, Software

PLATFORMS SUPPORTED PC/XT/AT, PS/2, Macintosh II, other industrial computers

GW Instruments
35 Medford Ave.
Somerville, MA 02143
(617) 625-4096

PRODUCT LINE Analog I/O, Digital I/O, GPIB Interface, Software

PLATFORMS SUPPORTED Macintosh Plus/SE/II

Hewlett-Packard
19310 Pruneridge Ave.
Cupertino, CA 95014
(303) 679-3279

PRODUCT LINE Analog I/O, Digital I/O, Signal Conditioning, GPIB Interface, Sensors

PLATFORMS SUPPORTED AT

Data Acquisition Hardware Manufacturers

Industrial Computer Design, Inc.
31355 Agoura Road
Westlake Village, CA 91361
(818) 889-3179

PRODUCT LINE Analog I/O, Digital I/O, Sensors
PLATFORMS SUPPORTED PC/XT/AT, PS/2

Industrial Computer Source
4837 Mercury Street
San Diego, CA 92111
(619) 279-0084

PRODUCT LINE Ruggedized PCs, Analog I/O, Digital I/O, RS-232/422/485 Interface
PLATFORMS SUPPORTED PC/XT/AT

IOtech Inc.
25971 Cannon Road
Cleveland, OH 44146
(216) 439-4091

PRODUCT LINE GPIB Interface, Analog I/O and Digital I/O via GPIB, Software
PLATFORMS SUPPORTED PC/XT/AT, PS/2, Macintosh Plus/SE/II, other industrial computers

Keithley Instruments
28775 Aurora Road
Cleveland, OH 44139
(216) 248-0400

PRODUCT LINE Analog I/O, Digital I/O, Signal Conditioning, GPIB Interface, Software
PLATFORMS SUPPORTED PC/XT/AT, PS/2, Macintosh Plus/SE/II

Keithley Metrabyte
440 Myles Standish Blvd.
Taunton, MA 02780
(508) 880-3000

PRODUCT LINE Analog I/O, Digital I/O, Signal Conditioning, GPIB and RS-232/485 Interface, Software
PLATFORMS SUPPORTED PC/XT/AT, PS/2, Macintosh Plus/SE/II

Microstar Laboratories
2863 152nd Ave. N.E.
Redmond, WA 98052
(206) 881-4286

PRODUCT LINE Analog I/O, Digital I/O, Software
PLATFORMS SUPPORTED PC/XT/AT

National Instruments
12109 Technology Blvd.
Austin, TX 78727
(512) 794-0100

PRODUCT LINE GPIB Interface, Analog I/O, Digital I/O, Software
PLATFORMS SUPPORTED PC/XT/AT, Macintosh SE/II, other industrial computers

Qua Tech
478 E. Exchange St.
Akron, OH 44304
(216) 434-3154

PRODUCT LINE GPIB and RS-232/422/485 Interface, Analog I/O, Digital I/O
PLATFORMS SUPPORTED PC/XT/AT, PS/2

Rapid Systems, Inc.
433 N. 34th St.
Seattle, WA 98103
(206) 547-8311

PRODUCT LINE Analog I/O, Digital I/O, PC-based Digital Oscilloscopes, FFT Analyzers, GPIB Interface, PCXI Hardware, Software
PLATFORMS SUPPORTED PC/XT/AT

Real Time Devices, Inc.
820 N. University Drive
P.O. Box 906
State College, PA 16804
(814) 234-8087

PRODUCT LINE Analog I/O, Digital I/O, Software
PLATFORMS SUPPORTED PC/XT/AT

Scientific Solutions, Inc.
6225 Cochran Road
Solon, OH 44139
(216) 349-4030

PRODUCT LINE Analog I/O, Digital I/O, GPIB Interface, Software
PLATFORMS SUPPORTED PC/XT/AT, PS/2, Macintosh II

Soltec
Sol Vista Park
12977 Arroyo St.
San Fernando, CA 91340
(818) 365-0800

PRODUCT LINE Analog I/O, Digital I/O, GPIB and RS-422/485 Interface, Software
PLATFORMS SUPPORTED PC/XT/AT

Strawberry Tree, Inc.
160 South Wolfe Road
Sunnyvale, CA 94086
(408) 736-8800

PRODUCT LINE Analog I/O, Digital I/O, Software
PLATFORMS SUPPORTED PC/XT/AT

Validyne Engineering Corp.
8626 Wilbur Ave.
Northridge, CA 91324
(818) 886-2057

PRODUCT LINE Analog I/O, Digital I/O, Signal Conditioning, Software

PLATFORMS SUPPORTED PC/XT/AT

World Precision Instruments, Inc.
375 Quinnipiac Ave.
New Haven, CT 06513
(203) 469-8281

PRODUCT LINE Analog I/O, Software

PLATFORMS SUPPORTED Macintosh

Ziatech Corp.
3433 Roberto Court
San Luis Obispo, CA 93401
(805) 541-0488

PRODUCT LINE Digital I/O, GPIB Interface, Software

PLATFORMS SUPPORTED PC/XT/AT

APPENDIX **B**

Data Acquisition Software Manufacturers

This appendix contains listings for manufacturers of data acquisition, display, and analysis software usable on personal computers. Hardware manufacturers who also supply software products are listed in Appendix A. Under the listing of PLATFORMS SUPPORTED, PC/XT/AT refers to both IBM and compatible systems. PS/2 refers to Micro Channel systems. MS DOS is implied for both.

Aptech Systems, Inc. 26250 196th Place SE Kent, WA 98042 (206) 631-6679	PRODUCT LINE PLATFORMS SUPPORTED	GAUSS (Data Analysis and Display) PC/XT/AT, PS/2
Binary Engineering 400 Fifth Ave. Waltham, MA 02154 (617) 290-5900	PRODUCT LINE PLATFORMS SUPPORTED	TECH*GRAPH*PAD (Data Analysis and Display) PC/XT/AT, PS/2, other industrial computers
DSP Development Corp. One Kendall Square Cambridge, MA 02139 (617) 577-1133	PRODUCT LINE PLATFORMS SUPPORTED	DADISP (Data Analysis and Display) PC/XT/AT, PS/2, other industrial computers
HEM Data Corp. 17336 West 12 Mile Road Suite 200 Southfield, MI 48076 (313) 559-5607	PRODUCT LINE PLATFORMS SUPPORTED	SNAPSHOT STORAGE SCOPE (Data Acquisition), SNAP CALC, SNAP FFT (Data Analysis and Display) PC/XT/AT

Keithley Asyst Software Technologies, Inc. 100 Corporate Woods Rochester, NY 14623 (716) 272-0070	PRODUCT LINE PLATFORMS SUPPORTED	ASYST, ASYSTANT, EASYEST, VIEWDAC (Data Acquisition, Analysis and Display) PC/XT/AT, PS/2
Laboratory Technologies Corp. 400 Research Drive Wilmington, MA 01887 (508) 657-5400	PRODUCT LINE PLATFORMS SUPPORTED	LABTECH NOTEBOOK, LABTECH CONTROL, ICONview (Data Acquisition, Analysis and Display) PC/XT/AT, PS/2, Macintosh II, other industrial computers
MACE Inc. 2313 Center Ave. Madison, WI 53704 (608) 244-3331	PRODUCT LINE PLATFORMS SUPPORTED	Data Analysis Subroutine Libraries PC/XT/AT
The Math Works, Inc. 21 Eliot St. South Natick, MA 01760 (508) 653-1415	PRODUCT LINE PLATFORMS SUPPORTED	MATLAB (Data Analysis and Display) PC/XT/AT, Macintosh Plus/SE/II
Odesta Corp. 4084 Commercial Ave. Northbrook, IL 60062 (312) 498-5615	PRODUCT LINE PLATFORMS SUPPORTED	Data Desk (Data Analysis and Display) Macintosh
Prescience Corp. 939 Howard St. #104 San Francisco, CA 94103 (415) 543-2252	PRODUCT LINE PLATFORMS SUPPORTED	Theorist (Data Analysis and Display) Macintosh
Preston Scientific 805 E. Cerritos Ave. Anaheim, CA 92805 (714) 776-6400	PRODUCT LINE PLATFORMS SUPPORTED	Signalys (Data Acquisition, Analysis and Display) PC/XT/AT

Spiral Software
6 Perry St.
Suite 2
Brookline, MA 02146
(617) 739-1511

PRODUCT LINE Easy Plot (Data Analysis and Display)
PLATFORMS SUPPORTED PC/XT/AT

Talton/Louley Engineering
9550 Ridgehaven Court
San Dieto, CA 92123
(619) 565-6656

PRODUCT LINE RT-DAS (Data Acquisition and Display with Networking)
PLATFORMS SUPPORTED PC/XT/AT, PS/2

Trimetrix, Inc.
444 NE Ravenna Blvd.
Suite 210
Seattle, WA 98115
(206) 527-1801

PRODUCT LINE Axum (Data Analysis and Display)
PLATFORMS SUPPORTED PC/XT/AT, PS/2

Wavemetrics
P.O. Box 2088
Lake Oswego, OR 97035
(503) 635-8849

PRODUCT LINE IGOR (Data Analysis and Display)
PLATFORMS SUPPORTED Macintosh Plus/SE/II

Wolfram Research, Inc.
P.O. Box 6059
Champaign, IL 61826
(217) 398-0700

PRODUCT LINE Mathematica (Data Analysis and Display)
PLATFORMS SUPPORTED AT, PS/2, Macintosh Plus/SE/II, other industrial computers

Bibliography

Advanced Micro Devices. "Personal Computer Products Data Book." Advanced Micro Devices, Sunnyvale, California, 1989.
Analog Devices, Inc. "Analog-Digital Conversion Handbook." Prentice-Hall, Englewood Cliffs, New Jersey, 1986.
Apple Computer, Inc. "Designing Cards and Drivers for the Macintosh Family." Addison-Wesley, Reading, Massachusetts, 1990.
Biber, C., Ellin, S., Shenk E., and Stempeck, J. The Polaroid ultrasonic ranging system. *67th Audio Engineering Society Proceedings*, New York, 1980.
Burr-Brown Research Corp. "Operational Amplifiers Design and Applications." McGraw-Hill, New York, 1971.
Cappellini, V. (ed.). "Data Compression and Error Control Techniques with Applications." Academic Press, Orlando, Florida, 1985.
Cooper, D., and Clancy, M. "Oh! Pascal!." W. W. Norton and Co., New York, 1982.
Cornejo, C., and Lee, R. Comparing IBM's Micro Channel and Apple's NuBus. *Byte, Extra Edition—Inside the IBM PCs.* 1987.
Cress, P., Dirksen, P., and Graham, J. "FORTRAN IV with WATFOR and WATFIV." Prentice-Hall, Englewood Cliffs, New Jersey, 1970.
DeMarre, D., and Michaels, D. "Bioelectronic Measurements." Prentice-Hall, Englewood Cliffs, New Jersey, 1983.
Dettmann, T. "DOS Programmer's Reference." Que Corp., Carmel, Indiana, 1988.
Eggebrecht, L. "Interfacing to the IBM Personal Computer." Howard W. Sams and Co., Indianapolis, Indiana, 1986.
Franklin, M. "Using the IBM PC: Organization and Assembly Language Programming." CBS College Publishing, New York, 1984.
Higgins, R. "Digital Signal Processing in VLSI." Prentice-Hall, Englewood Cliffs, New Jersey, 1990.
Hordeski, M. "The Design of Microprocessor, Sensor and Control Systems." Reston Publishing Co., Reston, Virginia, 1985.
Intel Corp. "Microsystem Components Handbook." Intel Corp., Santa Clara, California, 1985.
IBM Corp. "IBM Personal System/2 Hardware Interface Technical Reference." IBM, 1988.
IBM Corp. "IBM Technical Reference, Disk Operating System." IBM, 1986.

IBM Corp. "IBM Technical Reference, Options and Adapters." IBM, Boca Raton, Florida, 1984.

IBM Corp. "IBM Technical Reference, Personal Computer." IBM, Boca Raton, Florida, 1984.

IBM Corp. "IBM Technical Reference, Personal Computer AT." IBM, Boca Raton, Florida, 1985.

Johnson, T. A comparison of MC68000 family processors. *Byte,* September 1986.

Jordan, L., and Churchill, B. "Communications and Networking for the IBM PC and Compatibles." Prentice-Hall, New York, 1987.

Kernighan, B., and Ritchie, D. "The C Programming Language." Prentice-Hall, Englewood Cliffs, New Jersey, 1978.

Marshall, T., and Potter, J. How the Macintosh II NuBus works. *Byte.* December 1988.

Mason, W. (ed.). "Physical Acoustics." Academic Press, New York, 1968.

Microsoft Corp. "Macro Assembler for the MS-DOS Operating System—Programmer's Guide." Microsoft Corp., Redmond, Washington, 1987.

Microsoft Corp. "Microsoft C for the MS-DOS Operating System—Language Reference." Microsoft Corp., Redmond, Washington, 1987.

Motorola, Inc. "M68000 Family Reference." Motorola, Inc., Phoenix, Arizona, 1988.

Muratore, J., Carleton, H., and Austerlitz, H. Ultrasonic spectra of porous composites. *IEEE Ultrasonics Symposium Proceedings.* San Diego, California, 1982.

National Semiconductor Corp. "Linear Databook." National Semiconductor Corp., Santa Clara, California, 1982.

Norton, H. "Handbook of Transducers." Prentice-Hall, Englewood Cliffs, New Jersey, 1989.

Oppenheim, A., and Schafer, R. "Digital Signal Processing." Prentice-Hall, Englewood Cliffs, New Jersey, 1975.

Scalzo, F., and Hughes, R. "Elementary Computer-Assisted Statistics." Van Nostrand Rheinhold Co., 1978.

Solari, E. "AT Bus Design." Annabooks, San Diego, California, 1990.

Summer, S. "Electronic Sensing Controls." Chilton Book Co., Philadelphia, Pennsylvania, 1969.

Thomas, H. "Handbook of Biomedical Instrumentation and Measurement." Reston Publishing Co., Reston, Virginia, 1974.

Weber Systems, Inc. "C Language Users Handbook." Ballantine Books, New York, 1984.

Wells, P. Intel's 80386 architecture. *Byte, Extra Edition—Inside the IBM PCs.* 1986.

Index

Absolute optical encoder, 21–22
Absolute accuracy, 60–61
Accumulator, 98
Active filter, 32, 35
ADC, see Analog-to-digital converter
ADCs, see Analog-to-digital converters
Address clash, 69
Addressing, 65, 66, 74, 77
Aliasing, 59
Alphanumerics, 79
AM, see Amplitude modulation
AMD AM9513A, 204
Amplitude modulation (AM), 39
Analog circuit components, 25–31
Analog conditioning circuits, 31–39
Analog Devices, 214, 236, 295
Analog input card design, 91–92
Analog I/O, 201–203
Analog measurements, 6
Analog multiplier, 39
Analog signals, see Waveforms
Analog-to-digital converter (ADC), 2, 24, 30, 44, 91–92, 101–103, 201–203
  accuracy, 60–62
  characteristics, 58–62
  resolution, 58
  sampling or conversion rate, 58–59
Analog-to-digital converters (ADCs), 50–62, see also Analog-to-digital converter
  dual-slope, 53–54
  flash, 54–56
  ramp, 50–51
  servo, 51
  sigma-delta, 51–53
  special purpose, 62
  successive approximation, 56–58
  voltage-to-frequency (VFC), 54–56

Anti-aliasing filter, 59
Antilog amplifier, 38
Apple Macintosh, 4, 64–65, 81, 107, 127, 250, 293–294
Apple Macintosh II, 237–243, 247, see also Apple Macintosh
Application program, 97
ASCII, 114, 125, 134, 147, 152–153, 158, 213, 219, 222, 224, 228, 252, 253, 256
ASCII files, 148–150
ASCII-to-binary conversion, 152–153
Assembler, see Assembly language
Assembly language, 100, 101–102, 252, 254–257
Asynchronous communications, 127–128, 135–136
ASYST, 126, 220–223, 286, see also Keithley Asyst Software
ATE, see Automated test equipment
AT&T DSP32, 248
Autocorrelation, 280, see also Cross correlation
AUTOEXEC.BAT, 149, 292
Automated test equipment (ATE), 273

Backplane, see Passive backplane
Balanced line, 137, 140
Band pass filter, see Filters
Band reject filter, see Filters
BASIC, 78, 126, 188–191, 210–211, 253, 257–261, see also Microsoft QuickBASIC
BASIC (BAS) files, 149–150
Basic input/output system (BIOS), 70, 78, 80, 89, 95–96, 97–99, 250
  interrupts, 98–99, 117, 132
Batch (BAT) files, 147–149
Battery-backup CMOS RAM, 74

Index

BCD, *see* Binary coded decimal
Biased exponent, 177
Binary coded decimal (BCD), 173–175, 244
 instrumentation interface, 127
Binary codes, 41–43, *see also* Floating point formats
 fractional, 42–43, 176–177
 natural, 41–42, 172
 ones complement, 42
 twos complement, 42, 173
Binary Engineering, 227–228, 301
Binary point, 43
Binary resistance quad, 47
Binary Synchronous Communication (BSC), 141–142
BIOS, *see* Basic input/output system
Bisync, *see* Binary Synchronous Communication
Bit (Binary digit), 40–43
Bit compression ratio, 169
Bit-mapped graphics, 79–80
Bit weight, 41
Bits per second (BPS), 128
Bolometer, *see* Transducer, optical
Bonded strain gage, 13
Boot sector, 145
Bootstrap, 70, 96
BPS, *see* Bits per second
BSC, *see* Binary Synchronous Communication
Burr-Brown/Intelligent Instrumentation, 213–214, 236, 243, 296
Bus cycles, 83
Byte, 41

C programming language, 100, 103, 211, 252–254, 261–265, 271–272
Cache, 240
CAD, *see* Computer-aided design
Capacitive hygrometer elements, *see* Transducers, humidity
Carrier boards, 200, 213
CCD, *see* Transducers, optical
CD-ROM drives, 82
Central processing unit (CPU), 64, 99
 bandwidth, 101
Centronics interface, *see* Parallel interfaces
CFT, *see* Continuous Fourier transform
CGA, *see* Color Graphics Adapter

Circular average, *see* Running average
Clusters, 144–145
Color Graphics Adapter (CGA), 79
COM files, 150
Command-driven software, 217–218
Common mode noise, 137–138
Common mode rejection, 26
Comparator, *see* Voltage comparator
Compiler, 252–254
Compression ratio, 152
Computer-aided design (CAD), 80
CONFIG.SYS, 217
Continuous Fourier transform (CFT), 185
Convolution, 191–193
CPU, *see* Central processing unit
CRC, *see* Cyclic redundancy check
Cross correlation, 198
Current-to-voltage converter, 45
Curve fitting, 180–182
Cutoff frequency, 31
CYBEX extremity testing system, 287–292
Cyclic redundancy check (CRC), 142, 157
Cylinders, 143–144

DAC, *see* Digital-to-analog converter
DACs, *see* Digital-to-analog converters
DADiSP, 225–226, *see also* DSP Development Corp.
Daisy chain, 120
Dark count, 10
Data acquisition boards, 199–216, *see also* Appendix A
Data acquisition software, 216–228, *see also* Appendix B
Data analysis techniques, 178–198
Data communications equipment (DCE), 128
Data compression techniques, 151–170, 284
Data conversion, 44–45
Data streamer, 207, 220
Data terminal equipment (DTE), 128
Data Translation Inc., 211–213, 236, 241, 296
dB, *see* Decibels
DCE, *see* Data communications equipment
Decibels (dB), 41
Deconvolution, *see* Convolution
Delta encoding, 154–157

Delta Huffman encoding, 161, 284–285
DESQview, 107
Device drivers, 104
Device independence, 97
DFT, see Discrete Fourier transform
Difference amplifier, see Op amp circuits
Differential amplifier, 25–26
Differential signal, 137–138, 140
Differentiator, 34, see also Op amp circuits
Digital filter, 56, 193–194, 197–198
Digital I/O, 200–201
Digital I/O card design, 86–89, 92–94
Digital quantities, 40–45
Digital signal processing (DSP), 178, 185, 196, 197–198, 220, 247–248, 279
  boards, 247–248
Digitally controlled attenuator, 48
Digital-to-analog converter (DAC), 2, 44–45, 201–203
  characteristics, 48–49
Digital-to-analog converters (DACs), 44–49, see also Digital-to-analog converter
  fully decoded, 45
  multiplying, 48, 291
  weighted resistor, 46–47
Digitizing pad, 80
Direct memory access (DMA), 69–70, 71, 73, 74, 76, 89–90, 203, 219, 232
Directive, 255
Directories, 145–147
Discrete Fourier transform (DFT), 185–186
Disk operating system (DOS), 68, 71, 96–97, 99–100, 102, 104, 105–106, 107, 109, 222, 223, 240, 251, 252–253, 286
  disk structure, 143–147
  file types, 147–151
  interrupts, 99–100
  memory limitations, 107–110
Diskettes, 80–81
DMA, see Direct memory access
Doppler effect, 23
DOS, see Disk operating system
Dot-matrix printer, 82
Double precision, 176, 245
DRAM, see Dynamic RAM
DSP, see Digital signal processing
DSP Development Corp., 225–226, 301
DTE, see Data terminal equipment

Dual-slope ADC, see Analog-to-digital converters
Dynamic RAM (DRAM), 70, see also Random access memory
Dynamic range, 41, 153, 171–172, 245
Dynode, 9

EASYEST, 220, 223, see also Keithley Asyst Software
EBCDIC, 141
ECG, see Electrocardiogram
EGA, see Enhanced Graphics Adapter
EIA, see Electronic Industries Association
EISA, see Extended Industry Standard Architecture
Electrocardiogram (ECG), 281–282
  measurement system, 281–286
Electromagnetic interference (EMI), 233
Electronic Industries Association (EIA), 127
Embedded PC, 286
Embedded PC applications, 286–292
EMI, see Electromagnetic interference
EMM, see Expanded memory manager
Enhanced expanded memory, 110
Enhanced graphics adapter (EGA), 79
Enhanced small device interface (ESDI), 81
ESDI, see Enhanced small device interface
Euler's formula, 185
EXE files, 150–151
EXE2BIN, 151, 254
Expanded memory, 109–110
Expanded memory manager (EMM), 110
Extended Industry Standard Architecture (EISA), 231, 236–237, 293
Extended memory, 107, 109

Fast Fourier transform (FFT), 186–191, 192, 193, 194, 197, 222, 247, 255, 270, 279
FAT, see File allocation table
Feedback loop, 27
FFT, see Fast Fourier transform
File allocation table (FAT), 144–145
File server, 289
Filters, 25, 31–35
  band pass, 31–32, 34

band reject, 31–32, 34
digital, *see* Digital filter
high-pass, 25, 31–35
low-pass, 25, 31–35
Finite impulse response (FIR) filter, 197, *see also* Digital filter
FIR filter, *see* Finite impulse response filter
Firmware, 71, 95
First-order predictor (FOP), *see* Linear predictor
Flash ADC, *see* Analog-to-digital converters
Floating point formats, 176–178, *see also* Math coprocessors
Floppy disk drive, 80–81, 145
Floptical disk drive, 82
Flow chart, 270
FM, *see* Frequency modulation
FOP, *see* First-order predictor
FORMAT, 144
FORTH, 221, 254
FORTRAN, 265–267
Four-quadrant multiplying DAC, 48, *see also* Digital-to-analog converters
Fourier transforms, 184–191
Fractional binary, *see* Binary codes
Fragmented disk, 145
Frequency modulation, (FM), 30
Full duplex, 128
Fully decoded DAC, *see* Digital-to-analog converters
Function generator, 31
Function privacy, 262

Gage factor (GF), 13, 14
Gain bandwidth product, 27
Gas amplification factor, 9
Gas photodiode, *see* Transducers, optical
Gaussian distribution, 180
Geiger–Muller tube, 18
General purpose interface bus (GPIB), *see* Parallel interfaces
GF, *see* Gage factor
GPIB, *see* Parallel interfaces
Gray code, 22, 55

Hall effect, *see* Transducers, magnetic field

Hamming window, 195–196
Handshake, 112, 115, 117, 120, 121–122, 123, 129, 233
Hanning window, 195–196
Hard disk drive, 81, 145–146
Hardware interrupts, 87–89, *see also* Interrupts
Harmonics, 59
Harvard architecture, *see* Modified Harvard architecture
HDLC, *see* High-level data link control
Head meter, *see* Transducers, fluid flow
HEM Data Corp., 227, 301
Hercules Graphics Adapter (HGA), 79–80
Hewlett Packard, 119, 296
Hexadecimal notation, 66
HGA, *see* Hercules Graphics Adapter
High-level data link control (HDLC), 141
High-pass filter, *see* Filters
Hilbert transform, 196–197
Hot key, 105
HPIB, *see* Parallel interfaces, GPIB
Huffman encoding, 157–161

IBM Asynchronous Communications Adapter, 135
IBM Binary Synchronous Communications Adapter, 142
IBM DOS, *see* DOS
IBM Micro Channel Architecture (MCA), 63, 64, 231–236, 293
IBM OS/2, 106–107
IBM PC, 64, 105
IBM PC/AT, 64, 73–78, 92–94, 105, 106–107, 109, 110, 130, 135, 200, 213
IBM PC/XT, 64, 66–73, 83–90, 105, 110, 129, 200, 293
IBM PC/XT/AT computers, 4, 63–82, 105, 134, 143, 172, 199–200
IBM Printer Adapter Card, 118–119
IBM PS/2, 4, 64, 118, 231–237
IBM Synchronous Data Link Control Communications Adapter, 142
ICONview, 225, *see also* Laboratory Technologies Corp.
ICs, *see* Monolithic devices
IDE, *see* Integrated Drive Electronics
IEEE-488, *see* Parallel interfaces, GPIB
IEEE-488.1, 293, *see also* Parallel interfaces, GPIB

Index   311

IEEE-488.2, 293, *see also* Parallel interfaces, GPIB
IEEE-754 floating point standard, 176, 247
IFFT, *see* Fast Fourier transform
IIR filter, *see* Infinite impulse response filter
Ill-behaved software, 78
Incremental compiler, 254
Incremental optical encoder, 21
Index pulse, 21
Industrial PCs, 249
Industry Standard Architecture (ISA), 64, 73, 75–78, 92–94, 200, 213, 247, 249, 289, 293
Infinite impulse response (IIR) filter, 197–198, *see also* Digital filters
Integer formats, 171–175
Integer representations, 41–42
Integrated circuits (ICs), *see* Monolithic devices
Integrated Drive Electronics (IDE), 81
Integrator, 33, 53–54, 56, *see also* Op amp circuits
Intel 80286, 64, 73–75, 106, 226, 231, 235, 247, 255, 289, *see also* Intel 80x86 CPU family
Intel 80287, *see* Math coprocessor
Intel 80386, 64, 78, 106, 223, 226, 231, 247, 255, 293, *see also* Intel 80x86 CPU family
Intel 80386SX, 247, *see also* Intel 80x86 CPU family
Intel 80387, *see* Math coprocessor
Intel 80486, 64, 223, 226, 293, *see also* Intel 80x86 CPU family
Intel 8087, *see* Math coprocessor
Intel 8088, 64, 65, 66, 87, 106 *see also* Intel 80x86 CPU family
Intel 80x86 CPU family, 4, 63–65, 97, 105, 172, 252
Intel 8237, 89–90
Intel 8251A, 142
Intel 8255A, 201
Intel 8259, 87, 101
Intel 8273, 142
Interpolative encoding, 168–169
Interpolators, *see* Interpolative encoding
Interpreter, 252–253
Interrupt-driven software, 100–103
Interrupt levels, 97
Interrupt service routine, 89, 100–101

Interrupt vectors, 97
Interrupts, 67–69, 71–72, 74, 100–103, 232, *see also* Hardware interrupts; Software interrupts; Basic input/output system, interrupts; and Disk operating system, interrupts
Inverse FFT (IFFT), *see Fast* Fourier transform
Inverting amplifier, *see* Op amp circuits
I/O addressing, 66–68
I/O data transfers, 83–85
ISA, *see* Industry Standard Architecture
Isolation amplifier, 283

Josephson Junction, 16

Keithley Asyst Software, 220–223, 302
Keithley Instruments, 214–215, 297
Keithley Metrabyte, 206–211, 235–236, 241–242, 261, 297
Kernighan and Ritchie, 261, *see also* C programming language

Laboratory Technologies Corp., 223–225, 302
Labtech Control, 224–225, *see also* Laboratory Technologies Corp.
Labtech Notebook, 126, 223–225, 230, 243, *see also* Laboratory Technologies Corp.
LAN, *see* Local area network
Least significant bit (LSB), 41
Least squares fit, 180
LIM, *see* Expanded memory
Linear curve fit, *see* Linear regression
Linear interpolator, 168–169
Linear predictor, 166–168, 284
Linear regression, 181
Linear voltage differential transformer (LVDT), *see* Transducers, position or displacement
Linearity, 49, 62
LINK, 254
Linker, 253–254
Lithium chloride, 29
Local area network (LAN), 80, 289
Local curvature, 163–165
Log amplifier, 38–39

Longitudinal mode, 20
Lotus, 1-2-3, 219, 222, 224, 228
Low-pass filter, *see* Filters
LSB, *see* Least significant bit
LVDT, *see* Linear voltage differential transformer

Machine language, 251–252
Macintosh computers, *see* Apple Macintosh
Macro, 256–257
Marking level, 135
MASM, *see* Microsoft Macro Assembler
Mass storage devices, 80–82
Math coprocessor, 66, 78, 221, 224, 226, 240, 243–247
Math Works Inc., The, 227, 302
MATLAB, 227, *see also* Math Works Inc., The
MCA, *see* IBM Micro Channel Architecture
MCGA, *see* Multicolor Graphics Array
MDA, *see* Monochrome display adapter
Memory data transfers, 85–86
Memory models, 255
Memory refresh, 69, 70, 89, *see also* Dynamic RAM
Memory segmentation, 65–66
Menu-based software, 217–218
MFM, *see* Modified frequency modulation
Micro Channel, *see* IBM Micro Channel Architecture
Microprocessor, *see* Central processing unit
Microsoft, 71
Microsoft CodeView, 253
Microsoft Macro Assembler (MASM), 255–257
Microsoft QuickBASIC, 126, 261
Microsoft Windows, 107
Modem, 130–132, 152
Modified frequency modulation (MFM), 81
Modified Harvard architecture, 248
Modulation, 39
Modulator, 56
Monochrome display adapter (MDA), 79
Monolithic devices, 48
Monolithic temperature transducer, *see* Transducers, temperature

Monotonic, 45
Monotonicity, 49
Most significant bit (MSB), 41
Motherboard, 66, 70, 248
Motorola 68000 CPU family, 4, 64–65, 172, 237, 240, 241, 252
Motorola 68881, *see* Math coprocessors
Motorola 68882, *see* Math coprocessors
Motorola DSP56000, 248
Mouse, 80, 110–111, 238
MSB, *see* Most significant bit
MS DOS, *see* Disk operating system
Multicolor Graphics Array (MCGA), 79
Multi-drop, 140
Multifunction board, 204–205
Multiplexer (Mux), 3, 200, 245, 238, 274
Multiplying DAC, *see* DACs
Multi-tasking, 107, 110
Mux, *see* Multiplexer

National Instruments, 126, 215–216, 236, 243, 298
National Semiconductor ADC0808, 91
National Semiconductor INS8250, 135
Natural binary, *see* Binary codes
NMI, *see* Non-maskable interrupt
Non-DOS operating systems, 106–107
Noninteger formats, 175–178
Noninverting amplifier, *see* Op amp circuits
Nonmaskable interrupt (NMI), 67–68, 87, 245
Notch filter, *see* Filters, band reject
NuBus, 4, 64, 65, 237–243, 293
Null modem cable, 133
Numerical representation, 171–178
Nyquist frequency, 59
Nyquist theorem, 59

Object code, 252
OEM, *see* Original equipment manufacturer
Ones complement, *see* Binary codes
Op amp (Operational amplifier), 25–29, 46
Op amp circuits, 27–29
  difference amplifier, 28
  differentiator, 29
  integrator, 28

inverting amplifier, 27
noninverting amplifier, 28
voltage follower, 27
Op code, 252
Open architecture, 65
Open-collector driver, 120
Operational amplifier, *see* Op amp
Optical drives, 82
Optical interruptor, 291
Original equipment manufacturer (OEM), 273
OS/2, *see* IBM OS/2
Output offset, 54
Overlays, 107–108, 151
Oversampling, 56

Parallel Interfaces, 112–114, 114–127
  Centronics, 82, 114–119
  general purpose interface bus (GPIB), 119–127, 200, 215, 236, 243, 293
Parallel RLC circuits, 34–35
Parity bit, 136
Parity checking, 70
Partition table, 145–146
Pascal, 267–270
Pass band, 34
Passive backplane, 248–249
Passive filters, 32, 33–34
PC peripherals, 78–82
PCXI, 239
Peak detector, 37–38
Phase detector, 30
Phase-locked loop, 30
Photoconductive cells, *see* Transducers, optical
Photodarlington, *see* Transducers, optical
Photodiode, *see* Transducers, optical
Photoelectric effect, 8
Photomultiplier tube (PMT), *see* Transducers, optical
Phototransistor, *see* Transducers, optical
Photovoltaic cell, *see* Transducers, optical
Piezoelectric effect, 14
Piezoelectric transducers, *see* Transducers, force and pressure
Pipelining, 248
PLD, *see* Programmable logic device
Plotter, 82

PMT, *see* Transducers, optical
Polaroid, 274
Polled software, 100, 103
POS, *see* Programmable option select
Potentiometer, *see* Transducers, position or displacement
Power supply, 73, 76–77
Predictive encoding, 165–168
Predictors, *see* Predictive encoding
Printer, 82
Program segment prefix (PSP), 150
Programmable logic device (PLD), 86
Programmable option select (POS), 233
Pseudo-op code, 255
PSP, *see* Program segment prefix
Public labels, 102

Qua Tech, 129, 298
QuickBASIC, *see* Microsoft QuickBASIC

R-2R resistance ladder, 47–48
RAM, *see* Random access memory
Ramp ADC, *see* Analog-to-digital converters
Random access memory (RAM), 66
Rapid Systems, 249, 298
RC filter, 33–34
Read only memory (ROM), 66, 70–71, 78, 95
Reduced instruction set computer (RISC), 63
Resistance quad, *see* Binary resistance quad
Resistance temperature detector (RTD), *see* Transducers, temperature
Resistive hygrometer elements, *see* Transducers, humidity
Resolution, 58, 60
Resolution reduction, 153–154
Resonant frequency, 34
RISC, *see* Reduced instruction set computer
Ritchie, *see* Kernighan and Ritchie
RL filter, 34
RLL, *see* Run length limited
RMS distortion, *see* Root mean square distortion
ROM, *see* Read only memory

Root directory, 145–146
Root mean square (RMS) distortion, 163, 285
Rotational flowmeter, see Transducers, fluid flow
RS-232C, see Serial interfaces
RS-422A, see Serial interfaces
RS-423A, see Serial interfaces
RS-485, see Serial interfaces
RTD, see Transducers, temperature
Ruggedized PCs, see Industrial PCs
Run length limited (RLL), 81
Running average, 179

Sample and hold amplifier, 37, 62, 203
Sampling rate, see Analog-to-digital converter, sampling or conversion rate
Sampling reduction, 153–154
Scientific Solutions, 215, 236, 298
Scintillation counter, see Transducers, ionizing radiation
SCPI, see Standard Commands for Programmable Instrumentation
SCSI, see Small computer system interface
SDLC, see Synchronous Data Link Control
Sectors, 143–144
Segment:offset, see Memory segmentation
Semiconductor radiation detector, see Transducers, ionizing radiation
Semiconductor strain gage, 14, see also Transducers, force and pressure
Sensors, see Transducers
Serial interfaces, 112–114, 127–142
  RS-232C, 128–136, 200
  RS-422A, 137–139, 200
  RS-423A, 136
  RS-485, 140–141, 200
Series RLC circuit, 34
Servo ADC, see Analog-to-digital converters
Servo loop, 27
Settling time, 48–49
Sigma-delta ADC, see Analog-to-digital converters
Signal conditioning techniques, 24–25
Significant point extraction, 162–165
Single precision, 176, 245

Slew rate, 130
Sliding average, see Running average
Small computer system interface (SCSI), 81, 127, 237, 240
Snap-Calc, 227, see also HEM Data Corp.
Snap-FFT, 227, see also HEM Data Corp.
Snapshot Storage Scope, 227, see also HEM Data Corp.
Software documentation, 272
Software drivers, 125, 216–217
Software independence, 97
Software interrupts, 97–100, see also Interrupts
Software layers, 95–97
Source code, 253
Source level debugger, 270–271
Spacing level, 135
Special-purpose ADCs, see Analog-to-digital converters
Splines, 170
SQUID, see Transducers, magnetic field
Stack, 221, 256
Standard Commands for Programmable Instrumentation (SCPI), 293
Standard deviation, 179–180
Static RAM, 70
Statistical analysis, 179–180
STD bus, 248, 280
Strain, 13
Strain gage, see Transducers, force and pressure
Stress, 13
Subdirectories, 146–147
Successive approximation ADC, see Analog-to-digital converters
Superconducting quantum interference device (SQUID), see Transducers, magnetic field
Switched capacitor filter, 35, see also Filters; Digital filter
Symphony, 224, see also Lotus 1-2-3
Synchronous communications, 128, 141–142
Synchronous data link control (SDLC), 141

Tank circuit, 34
Tape drive, 81–82

Index 315

TECH*GRAPH*PAD, 227–228, *see also* Binary Engineering
Temperature regulation, 5
Temporary format, 177
Terminate and stay resident (TSR) programs, 102, 104–105
Texas Instruments TMS32010, 248
Texas Instruments TMS32020, 248
Texas Instruments TMS320C25, 214, 248
Thermistor, *see* Transducers, temperature
Thermocouple, *see* Transducers, temperature
Thermopile, *see* Transducers, optical; Transducers, temperature
Timer/counter, 66, 69–70, 74, 203–204
Tone decoder, 30
Torque, 289–290
Trackball, 80
Transducers, 1, 6–23
  fluid flow, 23
    head meter, 23
    rotational flowmeter, 23
    ultrasonic flowmeter, 23
  force and pressure, 12–15, 290
    piezoelectric, 14–15
    strain gage, 13–14
  humidity, 22
    capacitive hygrometer, 22
    resistive hygrometer, 22
  ionizing radiation, 17–19
    Geiger counter, 18
    scintillation counter, 19
    semiconductor, 18–19
  magnetic field, 15–17
    Hall effect, 16
    superconducting quantum interference device (SQUID), 16–17
  optical, 8–12
    bolometer, 12
    charge coupled device (CCD), 12
    gas photodiode, 9
    photoconductive cell, 10–11
    photodarlington, 11
    photodiode, 11
    photomultiplier tube (PMT), 9–10
    phototransistor, 11
    photovoltaic cell, 11
    thermopile, 7, 12
    vacuum photodiode, 8
  position or displacement, 19–22
    capacitive, 19
    inductive, 19
    linear voltage differential transformer (LVDT), 19–21
    optical encoder, 21–22, 290
    potentiometer, 19
    ultrasonic range finder, 22, 274–279
  temperature, 7–8
    monolithic, 8
    thermistor, 7
    thermocouple, 7
    thermopile, 7, 12
    resistance temperature detector (RTD), 8
Transfer function, 6, 32
Transistor–transistor logic (TTL), 40
Tri-state driver, 88
TSR, *see* Terminate and stay resident programs
TTL, *see* Transistor–transistor logic
Turnkey system, 273
Twos complement, *see* Binary codes
20 mA current loop interface, 129

UART, *see* Universal asynchronous receiver/transmitter
Ultrasonic flowmeter, *see* Transducers, fluid flow
Ultrasonic measurement system, 274–281
Ultrasonic range finder, 274–279, *see also* Transducers, position or displacement
Ultrasonic transducer, 14, 15, 274
Ultrasonic velocity, 278
Unadjusted error, 60
Unbonded strain gage, 13
Uninterruptible power source (UPS), 292
Unity gain frequency, 26–27
Universal asynchronous receiver/transmitter (UART), 135
Universal synchronous/asynchronous receiver/transmitter (USART), 142
Unix, 107
UPS, *see* Uninterruptible power source
USART, *see* Universal synchronous/asynchronous receiver/transmitter

Vacuum photodiode, *see* Transducers, optical

VCO, see Voltage-controlled oscillator
Venturi tube, 23
VFC, *see* Analog-to-digital converters
VGA, *see* Virtual Graphics Array
ViewDAC, 220, 223, *see also* Keithley Asyst Software
Virtual Graphics Array (VGA), 79
VME, 238, 248, 280
VMXI, 249
Voltage comparator, 29–30
Voltage-controlled oscillator (VCO), 30
Voltage follower, *see* Op amp circuits
Voltage-to-frequency converter (VFC), *see* Analog-to-digital converters

Wait state, 72, 90
Wait state generation, 90–91
Waveform processing, 182–184
Waveforms 1, 2
  continuous, 2
  monotonic, 2
Weighted resistor DAC, *see* Digital-to-analog converters
Weitek 1167, 246–247
Well-behaved software, 78
Wheatstone bridge, 35–37
Window functions, 195–196
Wirth, 267
Word, 41
Workstation, 63
WORM drive, 82
Wyse, 289

Xenix, *see* Unix
XON/XOFF, 157

Zero-order predictor (ZOP), 165–166, 285